Powder
Coating
Systems

Powder
Coating
Systems

William D. Lehr

McGraw-Hill, Inc.
New York   St. Louis   San Francisco   Auckland   Bogotá
Caracas   Hamburg   Lisbon   London   Madrid
Mexico   Milan   Montreal   New Delhi   Paris
San Juan   São Paulo   Singapore
Sydney   Tokyo   Toronto

Library of Congress Cataloging-in-Publication Data

Lehr, William D.
    Powder coating systems / William D. Lehr.
        p.   cm.
    ISBN 0-07-037072-9
    1. Plastic powders.   2. Plastic coating. I. Title.
TP1175.S6L44 1991
668.4'9—dc20   90-27853
                                                           CIP

CHAMELEON is a trademark of Thorid Electrostatic Powder Systems. ColorSPEEDER and FREEDOMCOATER are trademarks of Volstatic, Inc.

1 2 3 4 5 6 7 8 9 0   DOC/DOC   9 8 7 6 5 4 3 2 1

ISBN   0-07-037072-9

The sponsoring editor for this book was Robert W. Hauserman, the editing supervisor was Jim Halston, and the production supervisor was Pamela A. Pelton. It was set in Century Schoolbook by McGraw-Hill's Professional Publishing Book Group composition unit.

Printed and bound by R. R. Donnelley & Sons Company.

Marilyn
This venture and our adventure.
Thank you for helping me hold it together.
Bill

Contents

xii   Contents

Preface

I want you to read all of this book. There is a lot of common sense in it—and not a lot of extremely technical information. My object is not to teach you how to formulate and manufacture powder or how to build specific pieces of equipment, but rather to show you the anatomy of a good, well-operating powder coating system. I have intended it to be the definitive text on powder coating. But let me insert a cautionary note here: If you are planning to build your own system, don't waste your time reading this book. But if after it's built and doesn't work the way you wanted, then you can start over—and start by reading this volume. What you'll discover is that you'll have to use professionals who specialize in designing and building these systems if you are going to get what you really want.

This book tells things like they really are, the way they really happen in the field. It gives you the technical information you need, along with examples of situations as they happen in the real world. It does not give results and conclusions drawn from laboratory testing alone, nor is it filled with stories told after the storyteller has fitted you with rose-colored glasses.

You will get some good information from this book, given in an interesting fashion. After all, any good book, whether it's a spy novel, a cookbook, a book about automobiles, or a book on powder coating systems, has to provide material that will keep your interest.

The occasional duplication of information is meant to impress upon you the importance of the things you'll need to know if you are going to become involved in the powder coating industry. This repetition results from the advice of a very good and very wise friend, who has a lot of experience in telling people technical things in such a way that they remember what he says. This friend constantly advises me to

1. Tell them what you're going to tell them.

2. Then tell them.

3. Then tell them what you've just told them.

I guess what he means is, make sure they get the important points and remember them. That's what this book is all about.

Permit me to close this Preface with an appropriate ode, **Lehr's Law**:

> God made the heavens and the earth.
> The earth cooled.
> The dinosaurs were born and lived upon the earth.
> The dinosaurs died and became oil.
> In the 60s powder coating was invented.
> In the 70s the Feds created OSHA and EPA.
> In 1973 the Arabs invented OPEC and ruled the world of oil.
> In the 80s, and 90s, when companies who finished their products found
>   the only things which bothered them were OPEC, OSHA, and EPA,
> They eliminated the problems these three caused them by finishing
>   with powder.

Acknowledgments

I owe my education in the finishing industry to many people. To include all of their names here would be a difficult task; but Norm Allen, Stu McLaughlin, Glen Swanson, Carl Bolf, Brad Gruss, Ted Tylman, Jim Guirl, Darryl Ulrich, and Steve Kiefer are some of the friends who have helped me through the years. My thanks to all of them.

I also want to thank Bob Hauserman of McGraw-Hill, who had faith in me. Although powder coating is not his "long suit," he seemed to know from instinct that his volume on powder coating was something McGraw-Hill needed in its collection of technical books in order to give people working within the industry a place to look for the everyday help they need.

Thanks to all of you.

Bill Lehr

The History of Powder Coating

I am not going to attempt to define powder coating for you. There are some eloquent people in this world who could easily do that, but not me. I'm not going to start there. The fact that you're reading this volume means you know what powder coating is and what it's used for. You want to know more. So let's start there.

Powder coating, as we know it, is a mere "babe in arms," as the expression goes. Even so, we are already into the second generation in the industry. Let me explain.

Historically speaking, during the latter portion of the 1950s thermoset-type powder materials were introduced. They consisted of a simple epoxy material. The end product was considered a functional, not a decorative, coating that gave excellent chemical resistance and/or excellent electrical insulation values. Film thicknesses ran about 8 mils (200 microns, or micrometers). Thermoplastic materials were already in use then, and the new thermoset materials were first applied using the fluid-bed method, the same process used for the application of thermoplastic materials. It was a bit awkward to apply materials using this method; for one thing, preheating was required, and for another mil-thickness problems arose with each change in the size, shape, and weight of the products being coated.

In order to get the general idea here, you need to understand the simple differences between thermoset and thermoplastic materials. *Thermoset* materials (which is what this volume is mainly about) are those powder materials that, when applied and heated to a curing temperature, melt, flow, and then cross-link chemically. This produces a finish material of higher molecular weight. Once cured, this material, if reheated, will not remelt or reflow. On the other hand, *thermoplastic* materials, when applied and heated to a curing temperature, melt, flow, and cure. If they are heated again, they will remelt and

1

reflow. (In Chap. 3 we'll talk again of the many differences between the two basic materials.)

In the early 1960s, the powder coating industry began its amazing growth.* Shell Oil was instrumental in the development of epoxy powder materials and various methods of manufacturing them. Then in 1962, the application of powder with an electrostatic charge was accomplished by the S.A.M.E.S. Company of France. Since then, the development of powder coating systems, by whatever name—*pintura de polvo (talco), systema de polvo, pulverbeschichtungssystem, pulverlackering* system, *jauhelakkausjarjestelma,* or *praskove lakovani* system— has been pure excitement for me.

But not immediately for everyone else, it seems. When it was first introduced in the United States, powder gained little attention. There was no need at that particular moment in history to replace the liquid and solvent coatings that were available and widely used. They were inexpensive and in good supply, since the Arab nations had not yet made their move within the petroleum industry. (Paint solvents are, remember, a by-product of the petroleum industry.) Nor was there yet any particular interest in protecting the environment. If you lived downwind from a company that decided to install a paint line, you could expect no sympathy from any legal body or any politicians. It was called progress, and no one was ready to jump on the environmental bandwagon.

Some few companies in the United States did get into the manufacturing and marketing of powder coating materials and application equipment. Some survived; others did not. In the meantime, across the Atlantic Ocean, in Great Britain and on the continent, industry prospered in the manufacture of powder and powder application equipment. Companies like GEMA and Volstatic eventually introduced this produce to North America, giving notice to the manufacturing segment here that although powder was new, its time had come. And the industrial manufacturing sector of the United States responded. Bombarded by a new public environmental awareness and constrained by new environmental legislation, industry began to acknowledge the benefits of the powder process.

Once established, the powder industry made phenomenal growth. Sales of powder coating materials have continued to set records every year, making gains of from 15 to 20 percent annually. Powder application equipment has done even better. The establishment of the Powder Coating Institute (PCI) in the 1980s assisted the industry greatly, giving us a ready channel of information and literature. Figure 1.1 il-

---

*Portions of this history of powder coating were taken from S. T. (Sid) Harris, *The Technology of Powder Coatings,* 1976.

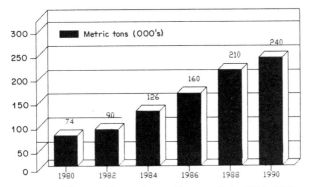

Figure 1.1 The world powder coating market, 1980–1990.
(*Courtesy of Powder Coating Institute*)

lustrates the growth of the world powder coating market between 1980 and 1990. Figure 1.2 shows the growth of North American powder sales between 1981 and 1989. On the basis of these statistics, the Powder Coating Institute estimates the growth from 1982 through 1995, shown in Fig. 1.3.

When an industry grows, so do those companies associated with the industry. For instance, powder coating lends itself to automatic spraying very well. The chart in Fig. 1.4 shows the sales of automatic powder coating equipment from 1981 through 1989. (Although there are thousands of manual guns in operation, no actual record has been kept of these units.) Powder coating clinics, seminars, tutorials, and workshops are frequently presented by the Society of Manufacturing Engineers (SME) and the Association of Finishing Processes (AFP), a

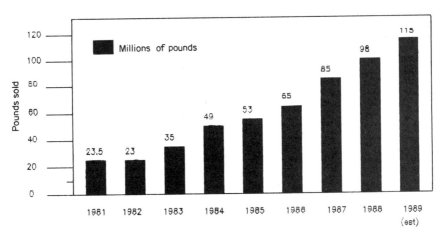

Figure 1.2 North American powder sales, 1981–1989. (*Courtesy of Powder Coating Institute*)

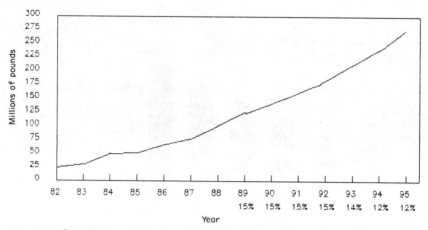

Figure 1.3 Growth of powder sales, 1982–1995. The rate of growth projected through 1995 does not reflect any major breakthroughs in the industry; it is based only on normal annual increases. (*Courtesy of Powder Coating Institute*)

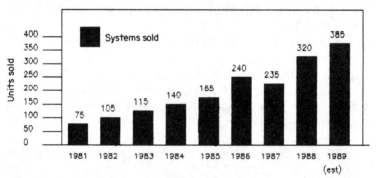

Figure 1.4 Automatic powder coating systems sold in North America, 1981–1989. (*Courtesy of Powder Coating Institute*)

section of SME, and similar sessions are presented by the Powder Coating Institute itself. Seminars, clinics, workshops, tutorials, and sessions, listed under other names but essentially devoted to powder coatings, have sprung up all over the country as well, indicating a very healthy interest in the subject.

Thus we can say that, with the impetus of interest during the late 1970s, powder really took off on its own in North America, and powder coating has now entered its second generation.

On a personal note, I would like you to know this book was written because of the efforts of people like Bob Lever, former managing director of Volstatic, and Norm Allen of Toronto, who got me interested in powder coating in the first place. Then there are distinguished people in the industry like Glen Swanson, Darryl Ulrich, Sal Lavano, Ron

Farrell, Chuck Danick, and many, many more who worked very hard to develop this industry. I want too to credit the efforts of Stu McLaughlin of Milwaukee, Wisconsin, from whom I have learned much through the years. Thanks to people like Stu, the powder industry has prospered. Then, of course, the industry has been helped through the work of Brad Gruss and the other people in the lab at Fremont Industries. Brad was the patient teacher from whom I learned the importance of proper pretreatment of metals before subjecting them to the powder coating process. That was one of the most important lessons I have learned: that it is the melding of this wonderful powder with a good pretreatment that gives the best finish possible on your products.

2

The Feasibility Study

Overview

The thesaurus always gives a good broad look at any word within its pages. For the word "feasible," it gives synonyms like possible, doable, practicable, workable, attainable; it even goes so far as to list advisable as a synonym. For words that stand in contrast to feasible, the thesaurus gives futile, hopeless, and impracticable. As an antonym, it offers impossible.

Now, a feasibility study would be a waste of time if you were trying to powder-coat lumber (excluding plywood), plastic substrates with low melting points, certain types of foods, and a few other things—not many other things, but a few. I have seen cured coatings of powder on paper and I know that, if you wanted to, you could powder-coat a bridge already erected and the hull of a ship. It is *possible* to coat all of these items, but it would only be *advisable* to coat some of them on a regular basis.

There are probably ways to powder-coat almost any product and cure it, but the bottom line is, Is the process practicable? I have seen some real miracles performed with powder; it really depends on what you're trying to accomplish. For you more practical people, the ones who want to coat metal products, powder is probably the answer for your company. If you are deluged with regulations regarding VOCs and with complex rules governing solvent emissions, you should take a close look at powder. You won't need to mire yourself and your company in an overwhelming search for the perfect coating; either powder will work or it won't. That can be easily established in a simple feasibility study.

Forming a Committee

To conduct a feasibility study of a powder coating system, regardless of the size of your company, form a committee. I have seen the strong leader of a small company make a fool of himself by taking complete charge of a new powder coating project. He made all of the decisions single-handedly because, after all, wasn't this his own company? The end result was the investment of thousands of dollars in an improperly finished product. On the other hand, the most successful systems I have ever seen were put together by an interested, efficient committee.

Try to include on the committee a person from every department that will be concerned with the finished coating on your substrate. Included on the committee should be at least one person involved in sales and marketing and at least one person from quality control. Sales and marketing people hear the competition bragging about the new miracle powder coating on their product and they begin to believe in this "smooth, continuous film that's impervious to everything, *forever!*" As a result, when and if you install a powder line, they'll expect a miracle. They'll want to subject your substrate product to unreasonable tests, and they'll come back to you dejected when the powder finish does not prove "impervious to everything, *forever.*" They'll not understand the impact-resistance and salt-spray tests. So get them involved at the beginning. Make certain they're present and that they understand the tests performed on cured powder films. Let them learn by being involved.

Keep quality control involved too. You'll face complete disaster if, a year after the powder system is started up, there are still conflicts over what powder can and cannot do.

Sales, marketing, and quality control should all be involved in the selection of the powder. They should know from testing what to expect from the new miracle, for if they know in advance what to expect, they will appreciate all the more the wonders of the cured film finish.

You should also have an independent consultant on the committee, one who doesn't sell equipment or materials, to advise you and to keep you on the right track. A consultant can clear up many questions a committee member may be afraid to ask.

Your committee should meet frequently. And its members, no matter what department they represent, should be responsible people, able to keep things moving and enthusiastic about the project.

Conducting the Tests

Keep the discussion and the tests reasonable. I have watched people literally destroy a product with a ball-peen hammer while trying to destroy its cured-powder surface film. When they were finally able to penetrate the coating, they gave a knowing laugh as if to say, "I knew I could do it," deriving personal satisfaction from having penetrated the coating to the bare metal and giving no thought to the fact that the product itself was no longer useful or workable and would have to be replaced. Match the finish to the product. If an impact resistance of 100 inch-pounds is all your product will ever need, then don't consider using powder that is rated at 160 inch-pounds and costs 40 cents a pound more than the material that has 100 inch-pounds impact resistance. Remember, too, that the powder coating doesn't need to be able to withstand 2000 hours of salt spray if the product is to be used in a kitchen.

Your committee should see tests of all sample parts and should get the results, good or bad. Comprehensive reports should be sent frequently to all involved.

Asking the Questions

Poor decisions at the beginning will give you a very strange powder system. Eventually, the tail-wagging-the-dog syndrome may take over, and the new system will begin to gobble up vast amounts of money just to be kept in operation. But the right powder coating system will be fairly simple and economical to operate if it is put together with some good intelligent testing. Your powder coating system can be the efficient, economical system your management envisioned when the subject of powder was first discussed.

Is powder really the answer for your company? It won't take long to make this decision.

What are your present specifications? What do you demand of your present coating? Your quality control people can tell you that if you ask them.

Can powder give you what you want in a cured film? If you need more than powder can offer, maybe you should investigate another process.

Are you going to integrate powder into an existing wet system? If you are, how much downtime will it take to mechanically integrate the powder equipment into the wet system?

What are the estimates of the cost of the new system? How about the installation and new utilities required?

Using Cost-Comparison Analysis

There are a variety of cost-comparison sheets that will enable you to compare your present wet-finishing-system operating costs with the projected operating costs of a powder system. Many powder companies and many powder equipment companies offer these free.

A simple analysis, using the chart shown in Fig. 2.1, will give you some idea of the advantages of using powder. Any representative of any powder manufacturer can give you the costs and the other information required for the form.

#	PROBLEM	SOLUTION	EXAMPLE	POWDER	OTHER
1.1	What is your coating cost per gallon/pound?	Insert the cost per gallon of wet coatings, and the cost per pound of powder.	$8.50		
1.2	The volume of solids in your coating?	This is the percentage of solids by volume as you receive your material.	43%	100%	
1.3	What is the cost per gallon of reducing solvent?	Insert the per gallon cost.	$22.25		
1.4	What is your coating solvent ratio?	IE: 5 gallons of paint + 2 gallons of solvent = 5.2	5.2		
1.5	What is the blended coating cost?	IE: (5 gallons of paint @ $8.50) + (2 gallons of solvent @ $2.25) = 5 x $8.50 = $42.50 + 2 x $2.25 = $4.50. $42.50 + $4.50 + $47.00 / 7 gallons = $6.71.	$6.71		
1.6	What percentage of actual solids is in your mixed coating?	Multiply line 1.2 by the number of gallons of coatings used in the blend. Then divide by the total gallons (coating and solvent) in the mix. .43 x 5 =2.15 / 7 = .307	31%	100%	
1.7	Specific gravity of the powder you will be using?	Your powder supplier can give you this.			
1.8A	Coverage at 100% efficiency (wet coating)	Multiply line 1.6 by 1604. .31 x 1604 = 497 sq. ft.	497 sq. ft		
1.8B	Coverage at 100% efficiency (powder)	Divide 192.5 by line 1.7			
1.9	Cured film thickness?	Insert average mil thickness of material applied to your product.	1.0 mil		
1.10	Utilization efficiency?	Conventional air spray guns:25-30% Electrostatic hand spray:50-60% Disc or bells:80-90% Electrostatic powder:95-98%	60%		
1.11	Coverage at utilization efficiency?	Divide line 1.8 by line 1.9. Multiply answer by line 1.10. 497 sq ft / 1.0 mil + 497 x .60 = 298.2.	298 sq ft		
1.12	Applied cost of material per sq. ft.	Divide line 1.5 by line 1.11. $6.71 / 298 = $0.0225	$0.023		

Figure 2.1   Simple cost-analysis chart. This material cost sheet shows the way to arrive at a material cost per square foot with material applied at a 1-mil thickness.

Emphasizing the Benefits of the Committee Study

Let me reiterate how strongly I feel about establishing a committee of interested people who represent a cross section of your company employees. Not involving people whose work will be affected by a new type of coating will quickly spell disaster. New system start-up will invariably bring unexpected questions; having other people involved will eliminate many of these questions before they are asked.

During the break-in period your committee should help everyone in your plant get accustomed to this new process of powder coating. It is better to have knowledgeable people within your company helping you over the hurdles rather than setting up larger ones.

No project ever gets off the ground until the first step is taken. I hope you and your committee take that first step together and then continue to walk in harmony through the completion of the project. I wish you every success.

3

The Powder Itself

Overview

The powder used in powder finishing is a plastic polymer, which comes in all imaginable shapes, colors, particle sizes, and composition. This miraculous flourlike material has found its way into almost every industry, into your house, your yard, your car, the kids' toys, even into the hearts of our EPA people, and of all other people dedicated to a clean environment.

Yes, powder is really in its second generation. I can remember the beginnings, when industry people were so excited and powder companies were giving all their energies to developing new materials. Now your powder manufacturer is becoming more interested in obtaining new business and retaining valued customers. Manufacturers know that if you have a problem, they also have a problem. They must accentuate service to you and your company, even as they continue to develop new and better materials to conquer new markets.

New developments are taking place daily in laboratories of specialty companies all over the world. Powder manufacturers buy their raw materials from these labs which are constantly looking for new ways to improve what they sell to the powder manufacturers.

Figure 3.1 shows the basic makeup of a powder formula. The formulation itself is not too complex. It is the specific standards designated by your company and the specific standards of other companies that make the differences in powders. I would not venture a guess as to how many different epoxy formulas there are. At an industry conference I recently heard that one particular company alone had over 100 epoxy formulas. I question that. I would think that the true figure is probably far more than 100.

In any event, the result is that powder products are available to you in many varieties and in all the colors of the rainbow. At each end of

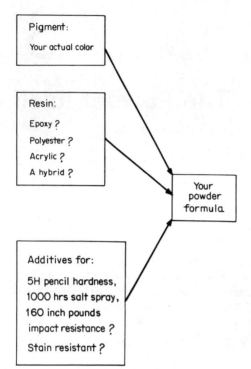

Figure 3.1 Basic makeup of a powder formula.

this rainbow are the specialty colors. Clear coatings that "don't yellow" are available in "water white." There are textures of all types achieved chemically and applied with one coat. There are veined coats of two colors, such as gold and black or silver and black. Then there is "candy apple red" and the other "high" or "electric" colors.

Let me share one of my favorite stories; it has to do with how the metal cabinets for computers were finished with wet coatings several years ago. I saw a finishing line that was composed of six massive, exhaust-belching monsters which spread the aroma of solvent for miles. These six spray booths were applying a very complex coating to the metal cabinets. The first two booths (one booth for each side of the product) applied the prime coat, which required 5 minutes of flash-off time before the parts entered a cure oven. After exiting the oven, the parts had to be cooled down for their next coating, which was applied in the next two booths. After a few minutes of flash-off time for this coating, called the base coat, the parts entered the last pair of booths, which were dedicated to the artistry of the industry in those days. Here the Michelangelos, the Rembrandts, the whoevers applied what was called the texture coat. Now I want you to know, this texture coat was a very precise kind of coat and there were sample panels hanging all around the artists applying the coatings. Woe to the one who interfered with the artist at work! Instantly, you could become infamous as the person who interrupted the artist's concentration, thus ruining thousands of dollars worth of parts in a single minute. The parts had to pass a rigid quality-control inspection; any 1-inch square of surface on which the spatter coat was applied had to have a specific quantity of spatter lumps, all within a specific size range. The reason for such exact color and texture matching was the fact that these finished cabinets would eventually be placed with cabinets finished in other plants, and everything had to match. If one "artist's" panels had twenty-five lumps or spatters per square inch, every other artist had to produce parts with twenty-five lumps or spatters per square inch. Can you imagine the logistics involved in doing this type of a project? Can you imagine the reject percentage and the cost per square foot to do the job?

I contrast this method with powder coating, which can be applied in a single coat to give a precise textured surface that is produced by a chemical reaction rather than by the hand of a very cantankerous "artist." Ten different powder coating shops located around the world can coat with the same powder, from the same batch, and cure at the same temperature, to get the same results, each using one single powder coating enclosure to do the job, instead of six, and not polluting the atmosphere with solvents. I believe this is what is called progress. And cost savings. I rest my case.

Thermosets and Thermoplastics

Through the years I have heard many presentations on powder coating, given all over the world. Some of these were about thermoset materials; some were about thermoplastic materials. If you are new to the industry and don't know the difference between the two, let me try to give you some idea so that when you're making decisions, you will know, up front, what is right for your particular job. I won't attempt to belabor each technical point, but I can give you enough information about the basic differences to keep you from making a mistake.

Table 3.1 outlines the differences for you.

Application methods

Thermoset materials are usually applied to a cold pretreated part using an electrostatic gun, either a hand gun, an automatic gun, or sometimes both. The application is done in an enclosure which has an

TABLE 3.1   A Comparison of Thermosets and Thermoplastics

Thermosets	Thermoplastics
Materials available:	
Epoxy	Polyethylene
Hybrids	Polypropylene
TGIC polyesters	PVC
Urethane polyesters	Nylon
Acrylics	Polyester
Application methods:	
Spray guns, either automatic or manual or a combination of both	Primarily a fluid bed, with the heated part immersed in the bed
Primers?	
Normally no	Usually yes, and usually liquid which requires curing prior to application of top coat
Film thickness	
Usually from 1.2 mils (30 microns)	From about 8 to 40 mils (250 to 1000 microns)

overspray powder recovery system attached. Thermoplastic parts are preheated, then dipped in a bed of fluidized powder. Some types of thermoplastic materials can be sprayed on.

Use of primers

Primers are normally not used in applying thermoset materials. The powder coating itself is considered adequate, though there are a few rare occasions when, for a very special reason, one may elect to use a primer or one of the special "zinc-rich" self-sacrificing materials available on the powder market. With thermoplastic materials, primers are used in most applications. The primers are usually liquid, and the parts are dipped or sprayed. The primers must then be cured before the top coating is applied.

Film thickness

The major difference between the two types of coatings is in their film thickness. Thermoset materials are used for thin-filmed decorative coatings, that is, to make the metal parts look pretty, to protect them from rusting, and—as many would have you know—"to better enhance your product." There is no question about it: Powders are simple to use, and they do an excellent job of protecting your finished product for a long time. The finishing cycle for thermoset materials is illustrated in Fig. 3.2.

Thermoplastic materials are normally used as a functional coating. As functional materials, thermoplastics have a very important and

Figure 3.2 The finishing cycle for thermoset materials.

specific job to do. For instance, a catcher's mask must have resiliency plus much strength. It needs a coating such as cannot be damaged by a baseball; when hit by a ball, the mask and the coating must spring back to the original shape. Dishwasher baskets must take the beating given by hot and cold water, by many types of detergents and surfactants, and, of course, by the heat of the drying cycle. These items, plus hundreds of others that do tough jobs, require the strength of a thermoplastic coating. The finishing cycle for thermoplastic materials is illustrated in Fig. 3.3. As you study the figure, remember that the cycle will vary slightly with the specific material.

When cured, thermoset materials cross-link chemically and produce a higher-molecular-weighted product. The important thing to remember is that, once cross-linked and cured, the thermosets will not reflow or remelt. On the other hand, when thermoplastic materials are cured and reheated, they will remelt, and reflow.

Now you know all you need to know about the differences between thermoset and thermoplastic materials. From this point on, we will be discussing only thermoset, or thin-film, decorative, coatings and their application. I'll let the other smart people of this world write the book on functional, or thermoplastic, materials.

Thermoset Powders

The thermoset family is, for my money, the more interesting. It takes limited skills to prepare and coat the product, and yet the finished parts can sit out there, being subject to all types of abuse for many years, and still look good. Two good examples of this are lawn mowers and snow blowers. Look at the abuse they receive constantly; yet their coating, a thin-film decorative coating, gives them good looks and protects them from the ravages of nature, chemicals, abuse, and your kids.

You'll be looking for many properties in the cured film of your product. You may want chemical resistance, salt-spray resistance, resistance to marring in packing and shipping, and resistance to sunshine.

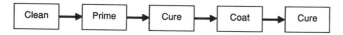

Figure 3.3 The finishing cycle for thermoplastic materials. Remember that the cycle will vary slightly with the specific material.

There are many qualities to be considered, and there are many powder products available from which to make your choice.

The basic families of thermoset powder comprise the following:

1. Epoxy

2. Hybrids

3. Urethane polyesters

4. TGIC polyesters

5. Acrylics

Epoxy

Epoxy materials are best suited to indoor use. When subjected to ultraviolet (UV) radiation of the sun, they tend to chalk and lose their gloss. Epoxy materials cure at temperatures well below 300°F; many will cure down in the 260°F range. The pencil-hardness range of epoxy can go to 7H, if needed. Its impact resistance is in the range of 160 inch-pounds, and epoxy can bend 180° around a 1/4-inch mandrel with no cracking or loss of adhesion. Epoxy materials have excellent humidity resistance and chemical resistance.

The epoxy materials have always been the workhorse of the industry. There are many important products that require epoxy materials, including kitchen cabinets and laboratory cabinets that require a certain type of chemical resistance. Epoxy-powder-coated parts under the hood of your car take abuse from the elements, and still perform well. I know of many epoxy-based products that are happily used outside when the loss of gloss from UV radiation is not a problem. When the U.S. Navy powder-coats exterior parts of vessels during refurbishing programs, they use epoxy for its superior salt-spray resistance, and really don't care about loss of the gloss, or chalking. In fact, if exterior parts are constantly cleaned or wiped off, they will not show the chalking typical of epoxy exposed to UV rays. The cleaning and wiping process will remove the chalking.

The general characteristics of epoxy powders are rated below:

- Hardness—excellent
- Flexibility—excellent
- Overbaking capabilities—fair
- Exterior durability—poor
- Corrosion protection—excellent
- Chemical resistance—excellent

Hybrids

Hybrid materials—combinations of epoxy and polyester—can give you the "best of both worlds," though the combination does not increase resistance to UV exposure. The technology was originally developed in Europe to yield the good properties of both products. People in the industry agree that the polyester resins in a hybrid material help slow down and reduce the yellowing characteristics sometimes associated with the overbaking of straight epoxy materials. The material seems to work extremely well when sprayed, thus increasing the odds for good transfer efficiency. The hybrid still remains, for all intents and purposes, an indoor material because of its UV vulnerability. The general characteristics of hybrid powders are rated below:

- Hardness—excellent
- Flexibility—excellent
- Overbaking capabilities—very good
- Exterior durability—poor
- Corrosion protection—excellent
- Chemical resistance—excellent

Urethane polyesters

Chemists within the industry, when talking about urethane polyester materials, will call them both urethanes and urethane polyesters. Actually, I guess one might define them as another form of a hybrid—hybrid for exterior use. One of the highlights of the urethane polyester coating is its ability to be applied at thin mil thicknesses. One to two mils (25 to 50 microns) are usually recommended. The following are general characteristics of urethane polyester materials:

- Hardness—very good
- Flexibility—very good
- Overbaking capabilities—very good
- Exterior durability—very good
- Corrosion protection—very good
- Chemical resistance—very good

TGIC polyesters

The initials stand for "triglycidal isocyranurate," a European-developed material. (I have heard that the initials also mean, "Thank goodness it covers.") TGIC polyester gives a superior coating, one with excellent exterior credentials. It is recommended that the TGIC variety materials be applied in a little heavier coating than the urethane polyester materials. Three to five mils (75 to 125 microns) do very well in maintaining all of the integrities of the film. The general characteristics of TGIC polyester materials are rated as follows:

- Hardness—excellent
- Flexibility—excellent
- Overbaking capability—excellent
- Exterior durability—excellent
- Corrosion protection—excellent
- Chemical resistance—very good

Acrylics

Acrylic materials are just becoming popular in the United States and are now being manufactured here. In the past, the technology and the materials have come from overseas. The general characteristics of acrylic materials are rated as follows:

- Hardness—very good
- Flexibility—fair
- Overbaking capability—good
- Exterior durability—very good
- Corrosion protection—fair
- Chemical resistance—very good

Table 3.2 summarizes the respective properties of the five powder families. Look over all of the available materials. Decide which can do the best job for your company. You probably will have a broad range of choices. If you are spraying more than one material, check to make certain of the complete compatibility of all the materials you plan to use.

Possible Problems to Be Considered

There are numerous other things, besides their general characteristics, that we need to look at when we discuss the various powders.

Orange peel

Orange peel is best when it is cooked and used in the making of marmalade; it also works well as a flavoring additive in gourmet dishes.

TABLE 3.2   A Comparison of the Powder Families

Property	Epoxy	Hybrid	Urethane polyester	TGIC polyester	Acrylics
Pencil hardness	H–7H	HB–3H	HB–3H	HB–H	2H–3H
Impact resistance, in-lbs	60–160	60–160	60–160	60–160	40–100
Gloss, 60°	3–100	10–100	15–95	20–90	10–90
Colors available	All	All	All	All	All
Clears	Yes		Yes	Yes	Yes
Textures	Yes	Yes	Yes	Yes	
Salt spray, hrs	1000 +	1000 +	1000 +	1000 +	1000 +
Condensing humidity	1000 +	1000 +	1000 +	1000 +	1000 +

But orange peel is not always so palatable in powder coatings and wet coatings. Some people would have you believe that the powder coating industry was the inventor of the orange-peel effect. Do me a favor. Look at cars parked on the street. What do you see? Orange peel! Look at your appliances. Orange peel! I have had a lot of experience with orange peel recently in the area of clear-coated polyester parts. My experience shows that if you stay with the manufacturer's suggested limits when applying the coating, there is usually very little, if any, orange peel effect. When you don't stay within the suggested limits, when you spray less (lighter mil thickness) or more (heavier mil thickness) than recommended, you are going to get a form of orange peel. So, at least with clear coatings, stay within the suggested coating range.

But there are many cases when manufacturers want and insist upon orange peel for their products. It hides a multitude of manufacturing sins. In appliances it looks better, reminding me of porcelain enamel. And it's easy to take care of from a cleaning standpoint. For instance, if you wipe down the refrigerator while wearing a ring with a sharp stone inset and the ring turns on your finger, the orange peel will "disguise" the resultant scratch.

The heavier film buildup of TGIC polyesters may show some orange peel. But they have been coating exterior wall panels of office buildings with TGIC for about 20 years now, and many of the buildings still have their original coating.

Remember, many coatings can be formulated to give you an orange-peel effect. On the other hand, proper manufacture, formulation, application, and cure can help reduce orange peel. The actual application of the material is up to you; if you apply the material properly, you can reduce orange peel.

Improper film thickness

Powder manufacturers go to great lengths to give you exactly what you want in the properties of a powder film. They can give you properties so new, they haven't even been advertised yet. But the one thing no one has yet learned how to do, is to make certain that you and your company use the powder the way it was designed to be used. Powder manufacturers will tell you what mil-thickness range you should use when spraying your powder and then pray (many times vainly) that you will spray within the limits they have suggested. They have many reasons for suggesting those limits, and you should heed them. If they tell you to spray 2.5 mils and you try to get by with a single mil, saying to yourself, "They just want to sell more powder," *you* will be the loser.

As a hypothetical example, let's say your particular powder was made to give you 1000 hours of salt spray and 160 inch-pounds of impact resistance, provided that you spray it at from 1.75 to 2.75 mils. Being conservative, you decide to spray at about 1.2 mils. How in the world is 1 mil of powder going to hold up to 160 inch-pounds of impact? What happens when someone scratches your 1-mil film and penetrates the coating to the bare surface? Because you wanted to save a couple of pennies, your finished product is going to prematurely lose its finish. You, your company, and the eventual user *all* end up the loser! The same situation occurs when you put on 5 mils and the suggested maximum thickness is 2.5 mils. The 160 inch-pound impact resistance drops literally like a rock; the nonstable thick coating will probably be very soft, *or* it could fracture at 80- to 100-pound impact.

You should also be aware that not staying within the suggested range will tend to increase the possibility of getting some orange peel.

Color contamination

Contamination is a defect that shows up on the finished part. The best way to describe it is to call it the "salt-and-pepper" effect. Say that you're spraying white, and then you decide it's time to spray black. You clean the system down, put the black powder into the system, and start to spray. Everything looks good to the eye during the spraying process, but when the parts come out of the oven, there you see it: the telltale salt-and-pepper effect of white powder contaminating black. It happens! Not all the time, but once is enough.

The problem comes from not thoroughly cleaning the system between color changes. A thorough cleaning of the application and reclaim system will eliminate the contamination. When you change colors the first few times in a new system, check very carefully for color contamination. Usually its not necessary to point out the problem to the people working on your finishing line more than once—they won't want to repeat the mistake either.

Contamination from incompatible powders

Your company may be one of those that sprays epoxy all of the time. But once in a while, maybe twice a year or so, it becomes necessary for you to change from the epoxy to a polyester. Believe me, it is impossible to manufacture a hybrid powder within your system. Epoxy just won't mix with a polyester that way. Instead, you'll get a very unusual finish, one you'll recognize immediately. Your line people will also easily recognize the incompatibility of the two products, as they try to melt, flow, and cure together. It will remind you of the reaction that occurs when you try to mix oil and water.

The same problem will occur when you try to use some types of acrylics with other materials. Sometimes they don't mix at all. Like the epoxy-polyester combination, acrylics and other generic materials will sometimes contaminate each other. Consult your supplier before making any line changes.

Silicone contamination

Silicones and powder do not mix. If you have been in finishing for any length of time, or have worked with wet coatings, you have probably experienced the problems that occur when any amount of silicone enters your finishing system.

Silicone contamination is easy to see on your finished parts. A small crater of cured powder will encircle the silicone; it appears to the eye as if the curing powder has attempted to separate itself from the particle of silicone.

Silicones have a habit of "just showing up." But with any kind of detective skills you can trace the silicone to its source. Metal fabricators sometimes put silicone oils on moving parts of presses and other manufacturing equipment. Occasionally an equipment operator will discuss a lubricating problem with a sales engineer who might recommend a "new lubricant" to cure the problem. The only problem is it contains silicones. Silicone oils do a fantastic lubricating job, but don't use them if you have a finishing line.

The silicone source can sometimes be especially baffling. In one instance, we finally located the source by observing that the marks occurred where the parts were held when being hung on the conveyor line; we then traced the silicone to a special hand cream used by the workers to soften chapped hands.

My first experience with silicone contamination occurred when, after searching for the silicone source for about a week, we were able to trace it to the plant heating system, where the fresh air intake was bringing in silicones expelled from the plastics manufacturing plant in the building next door, which used silicones as a mold release on certain dies. Fortunately for my client, the plastic plant soon moved to larger quarters.

Whenever you get an unexplained contamination, it's good to get advice from the technical service representative of your powder manufacturer.

Quality Control in Shipping and Storing Your Powder

When your purchasing department buys powder, you know when you bought it, when it arrived at your plant, and when it was paid for. But what else do you know about it, except for the fact it's sitting there on a pallet? Well, the numbers and letters listed on powder cartons tell much about your powder inventory. Learn what this information means and log it in a book. When you use powder, it should be on a "first in, first out" basis. New powder should be sampled, and a check should be made as to how close it is to your original sample, and also how close it is to a cured sample from the last batch.

By running an immediate sample you can determine whether or not your powder was, in fact, shipped in a refrigerated truck as you specified and paid for. You will quickly learn whether the refrigerator may have been shut off when the truck was parked in a freight terminal for the weekend. Some powders do not require refrigeration when being transported, the powder manufacturers tell me. But most powders are made to be stored in atmospheres of 80°F or lower, and the western desert areas can get hot, warming the truck interior. I have personally had powder agglomerate while sitting at a border crossing between the United States and Mexico.

Other disasters can occur from the time your powder leaves the manufacturing plant until it arrives at your plant. You and your powder supplier need to know immediately if some problem does occur. Coat some samples as a test. After you have had the powder for a few months, it will be difficult to do much tracing on a bad batch of powder, especially if it was ruined by some factor neither you nor the powder manufacturer had any control over. Keep the sample panels. Mark them and log the batch numbers on the shipment. Again, make certain you use the "first in, first out" method to rotate your powder stock.

Powder should not be stored in warm temperatures. How warm is warm? Powders vary so much, it is hard to give you an exact temperature. Most manufacturers will agree on 80°F as a maximum temperature. Powder suppliers can tell you if you need refrigerated storage. Don't store your powder on top of your oven or on a pallet close to the roof of your building. Either location will give you a solid mass of cured plastic in the powder cartons.

If your powder storage area needs to be kept cool, consider an air-conditioned room for your application equipment and store the powder in that room. If you don't and can't have an air-conditioned application room, build a small room of 2-by-4's and place a room air conditioner in one wall. Make sure the air conditioner is operating all of the time.

Powder works best when it is acclimated to the temperature in and around your powder application equipment. So it's a good idea to place the powder to be used on a given day close to the powder feed hopper. Open the carton, but not the plastic bag. This will help bring the powder to the temperature in the area.

Sieving Your Powder

Find out from your manufacturer if your powder needs to be sieved. If it does, find out when it should be sieved. Should it be sieved as a virgin powder? When reclaimed and mixed with virgin? At both times?

Have the sieve built into your application system if possible. It makes less work for your employees, gives you less spillage, and affords less chance of contamination. If the sieving device is built into your application system, it will give you more control over the destiny of your powder, and one less potential problem to worry about. There will be more information given on sieving in Chap. 9, which is devoted to ancillary equipment.

Tests, Tests, and More Tests

In any industry, there has to be a benchmark, a set of standards or rules by which laypeople and technicians can judge and decide the merits of a given product. Within the powder coating industry most standards are set by the American Society of Testing Materials (ASTM), which conducts tests on cured films. Your powder company can show you how these tests are performed. I have taken the liberty here of listing some of the popular tests you'll hear about (Table 3.3). A complete up-to-date list of the various tests is available from ASTM.

Usually, but not always, tests are made on metal panels which have been purchased by the powder manufacturer but pretreated by the panel manufacturer. Some of these panels in themselves are standards within the industry.

Your company will normally pretreat metal prior to the application of a powder coating. Hence running tests on a nonpretreated panel

TABLE 3.3 Common ASTM Tests for Powders

Test	Test number	Value/results
Specific gravity	ASTM D792	1.2 to 1.8
Gloss (60°)	ASTM D523	Will vary with product
Impact resistance	ASTM D2794	Up to 160 in-lbs
Flexibility	ASTM D1737	Conical mandrel testing
Flexibility	ASTM D522	
Pencil hardness	ASTM D3363	Measured in units of B or H
Crosshatch adhesion	ASTM D3359 (method B)	No lift of film
Salt-spray resistance	ASTM B117	Measured in hours
Humidity resistance	ASTM D2247	Measured in hours

will not give valid results. By the same token, all tests performed on behalf of your company should be performed on the same type of steel you use that has been through the same pretreatment system you will be using in production.

Setting Up Your Powder Formula

Your exact needs can be met by the powder formulators. You may need some specific qualities in your coating, or you may not care about specific requirements. The choice is yours.

Most likely, you have a set of standards for your product. Show these standards to your potential powder suppliers, who can then establish the exact formula you'll need. If you don't have a set of standards, you should develop one. You must decide what you will need for your product. I would advise you to be very realistic about your requirements.

As a hypothetical example, if you make an electric can opener for the kitchen of the average house, you will probably want to use an epoxy material. That eliminates all of the polyester materials. You may even give some consideration to a hybrid. How about salt-spray resistance? There's no reason to want 1000 hours salt-spray resistance, so why put it into the formula? On the other hand, the can opener will need a hard surface so that when it comes time to remove kitchen stains with strong detergents and slightly abrasive materials, the surface will remain in place. One doesn't normally bend an electric can opener, so there isn't much need for a lot of flexibility of the surface.

Another way to look at it is this way. If a 100 inch-pound blow will break your zinc die-cast part, why pay for powder that has a 160 inch-pound impact resistance? Beware. When you are too cautious or too profligate in one direction, you may introduce a problem in the very property you want to protect. And remember that the cost per pound of the powder will be based somewhat on the formula the powder company eventually puts together for you.

Calculating Your Powder Costs

There are many formulas by which you can calculate your per square foot cost of powder. As a matter of convenience, I've included one here for you. (See Fig. 3.4.) Most powder manufacturers, and equipment manufacturers, have charts or small calculating devices they will give you if you are interested in monitoring your material costs.

Cure Cycles of Various Powders

Powders usually have an extensive curing range, and there'll be a lot of flexibility in the cure cycles. This can help you in many ways, but there are a few precautions to be taken. Discuss the matter with your powder supplier if it appears you will have a problem with the suggested cure cycle.

SPECIFIC GRAVITY

MIL THICKNESS	1.0 s/g	1.1 s/g	1.2 s/g	1.3 s/g	1.4 s/g	1.5 s/g	1.6 s/g	1.7 s/g	1.8 s/g	1.9 s/g	2.0 s/g
1 mil	193.2	175.6	161.0	148.6	138.0	128.8	120.8	113.6	107.3	101.7	96.6
2 mils	96.6	87.8	80.5	74.3	69.0	64.4	60.4	56.8	53.7	50.9	48.3
3 mils	64.4	58.5	53.7	49.5	46.0	42.9	40.3	37.9	35.8	33.9	32.2
4 mils	48.3	43.9	40.3	37.2	34.5	32.2	30.2	28.4	26.8	25.4	24.2
5 mils	38.0	36.1	32.2	29.7	27.0	26.8	24.2	22.7	21.8	20.3	19.3
6 mils	32.2	29.3	26.8	24.8	23.0	21.5	20.1	18.9	17.9	17.0	16.1
7 mils	27.6	25.1	23.0	21.2	19.7	18.4	17.3	16.2	15.3	14.5	13.8
8 mils	24.2	22.0	20.1	18.6	17.3	16.1	15.1	14.2	13.4	12.7	12.1
9 mils	21.5	19.5	17.9	16.5	15.3	14.3	13.4	12.6	11.9	11.3	10.7
10 mils	19.3	17.6	16.1	14.9	13.8	12.9	12.1	11.4	10.7	10.2	9.7

To arrive at a per square foot cost per pound of powder, use the following formula and calculation:
A. powder cost per pound____ B. Specific gravity____ C. Dry film thickness____
D. powder volatiles 1%   E. collection efficiency 98%

Theoretical coverage per pound: $\dfrac{192}{B}$ = _F_ square foot/ mil

Actual coverage at applied thickness: $\dfrac{B \times (1.00-D) \times F}{C}$ =

$\dfrac{A}{G}$ = $0. per square foot cost

Figure 3.4 Cost calculation for a powder.

In looking over specification sheets put out by powder manufacturers, you'll see an especially wide range for epoxy materials. I know some epoxies cure as low as 250°F. It will take about 25 minutes for this material to cure. There are other epoxies which can be cured in the 400 and 450°F range. Their time cycles will be much shorter, something in the 5- to 12-minute range. And then there are very special epoxies which cure in much less than 5 minutes.

Two other things to consider: Remember to factor in the length of time it takes to raise your part to the curing temperature, and remember that it's necessary for some powders to be brought to *at least* a specific temperature, and held at that temperature for x minutes. You can go above the specific temperature, but not below.

You'll find a lot more on oven curing, oven cycling, and part temperatures in Chap. 7.

Determining Chemical Resistance

Frequently we are asked about the chemical resistance of particular powders. Figure 3.5 shows a chart that one powder manufacturer uses in answering these common questions about that one property of a powder.

Special Powder Coatings

Special coatings are available for special applications. Many industries require formulations for specific uses, and many items need protection from specific chemicals or other elements. As an example, there are pipe coatings made for the oil-drilling industry. These products must meet rigid specifications with regard to their ability to resist rapid oxidation from the natural elements found below the surface of our earth.

There are special textured finishes, and there are wrinkled finishes. These finishes are applied in one coat, receiving their texture or wrinkle from a chemical reaction.

Specific powders are available for "in-mold" coating. I first saw samples of this in Great Britain several years ago. An open mold is sprayed with powder, then a sheet of plastic is inserted. The two parts of the mold are brought together. The cured powder results in an abrasion-resistant, attractive coating for shower stalls, bathtubs, and lavatories. Similar materials are made for the automotive industry. Electrostatic-grade nylon materials are also available. The recommended mil thicknesses for these materials range from 3 to 6 mils.

Rebar, as the reinforcing cable and/or rods used below the surface of concrete highways, bridge decks, and other concrete structural mem-

Chemical	Epoxy		Vinyl		Nylon		Polyester (Thermoplastic)		Polyester (Thermoset)	
	C	H	C	H	C	H	C	H	C	H
Acids:										
Acetic, 10%	F	N	F	N	F	P	F	P	F	P
Acetic Glacial	N	N	N	N	N	N	P	P	P	P
Benzene Sulfonic, 10%	E	E	E	E	F	P	F	P	F	P
Benzoic	E	E	E	E	P	P	F	F	E	E
Boric	E	E	E	E	G	F	G	G	E	E
Butyric, 100%	P	N	G	N	G	F	F	P	F	P
Chloracetic, 10%	E	E	F	N	P	P	G	F	E	E
Chromic, 5%	F	N	E	G	G	F	F	F	P	P
Citric, 10%	E	N	E	E	G	F	G	G	E	E
Fatty Acids	E	E	E	F	G	G	G	G	E	E
Fluosilicic	N	N	E	E	P	P	P	P	P	P
Formic, 90%	E	F	F	N	P	P	P	P	P	P
Hydrobromic, 20%	G	G	E	E	P	G	G	G	G	F
Hydrochloric, 20%	E	G	E	E	P	P	G	G	G	F
Hydrocyanic	E	E	E	E	F	P	G	G	E	E
Hydrofluoric, 20%	G	G	E	E	F	F	F	F	P	P
Hypochlorous, 5%	F	N	E	E	F	P	F	P	G	F
Lactic, 5%	F	N	E	E	E	E	G	G	F	P
Maleic, 25%	E	E	G	F	F	F	G	G	E	E
Nitric, 5%	E	G	E	E	P	P	F	F	F	F
Nitric, 30%	G	P	E	F	P	P	P	P	P	P
Oleic	E	E	E	F	E	E	E	E	E	E
Oxalic	E	E	E	E	F	F	G	G	E	E
Phosphoric	G	F	E	E	P	P	G	F	G	F
Picric	G	F	P	N	G	G	G	F	G	F
Stearic	E	E	E	F	E	E	E	E	E	E
Sulfuric, 50%	G	F	E	E	P	P	P	P	F	P
Sulfuric, 80%	F	N	E	G	P	P	N	P	N	N
Tannic	E	E	E	G	G	G	G	G	E	E
Alkalies:										
Ammonium Hydroxide	E	G	E	E	F	F	P	P	P	P
Calcium Hydroxide	E	E	E	E	F	F	P	P	P	P
Potassium Hydroxide	E	E	E	E	F	F	P	P	P	P
Sodium Hydroxide	E	E	P	P	F	F	P	P	P	P
Acid Salts:										
Aluminum sulfate	E	E	E	E	E	E	E	E	E	E
Ammonium Chloride*	E	E	E	E	E	F	E	E	E	E

Chemical	Epoxy		Vinyl		Nylon		Polyester (Thermoplastic)		Polyester (Thermoset)	
	C	H	C	H	C	H	C	H	C	H
Acid Salts (cont.):										
Copper Chloride*	E	E	E	E	E	G	E	E	E	E
Iron Chloride*	E	E	E	E	E	G	E	E	E	E
Nickel Chloride*	E	E	E	E	E	G	E	E	E	E
Zinc Chloride*	E	E	E	E	E	F	E	E	E	E
Alkaline Salts:										
Barium Sulfide	E	E	E	E	G	F	E	E	E	E
Sodium Bicarbonate	E	E	E	G	F	E	E	E	E	E
Sodium Carbonate	E	E	E	G	F	E	E	F	E	F
Sodium Sulfide	E	E	E	G	G	F	E	F	E	F
Trisodium Phosphate	E	E	E	E	G	F	F	P	G	F
Neutral Salts:										
Calcium Chloride*	E	E	E	E	G	F	E	E	E	E
Magnesium Chloride*	E	E	E	E	E	E	E	E	E	E
Potassium Chloride*	E	E	E	E	E	E	E	E	E	E
Sodium Chloride*	E	E	E	E	E	E	E	E	E	E
Solvents:										
Alcohols	E	E	E	E	E	G	E	E	E	E
Aliphatic Hydrocarbons	E	E	F	P	E	G	G	G	G	G
Aromatic Hydrocarbons	E	E	P	N	F	P	P	P	G	F
Chlorinated Hydrocarbons	F	F	N	N	E	G	P	N	P	N
Ketones	F	F	N	N	E	F	P	P	F	P
Ethers	F	F	N	N	E	E	F	P	F	P
Esters	F	F	N	N	E	F	P	P	F	P
Gasoline	E	E	E	F	E	E	G	G	E	E
Carbon Tetrachloride	E	E	N	N	E	E	G	G	G	G
Organics:										
Aniline	G	P	N	N	F	P	P	P	P	P
Benzene	E	E	P	N	E	E	P	P	F	P
Formaldehyde, 37%	E	G	E	E	G	F	G	F	G	G
Phenol, 5%	G	F	N	N	P	P	G	F	G	F
Mineral Oils	E	E	P	P	E	E	E	E	E	E
Vegetable Oils	E	E	F	F	E	E	E	E	E	E
Chlorobenzene	G	P	N	N	E	E	P	P	G	F

KEY: E-no attack
G-appreciably no attack
F-some attack, but useable in some instances
P-attacked, not recommended for use
N-rapidly attacked
C-cold, 70°F.
H-hot, 180°F. or boiling point of solvent
*-and nitrate and sulfate

These data are intended only as a preliminary guide for material selection. Final selection should be made after consulting with MORTON POWDER COATINGS and after testing for your specific requirements.

MORTON
POWDER
COATINGS

Think Powder. Think Morton.

Morton Thiokol, Inc., Morton Chemical Division, Powder Coatings Group
P.O. Box 15240, Reading, PA 19612-5240 Phone (215)775-6600
MPCL 10/88-019

Figure 3.5 Chemical resistance of typical powder coatings. (*Courtesy of Morton Powder Coatings*)

bers exposed to marine and/or salt environments is commonly known, must have a special coating. The electrical industry has used special epoxies designed for specific dielectric properties for many years. Release coatings or nonstick-type coatings, for kitchen appliances are also available; they too have been in use for several years. And fluorescent coating systems are available in several forms, in a full range of colors.

Architectural Aluminum Coatings

For the past couple of years in the United States there has been much talk about architectural aluminum powder coatings and their ability to meet the rigid standards in place within the building industry. I have listened to many conversations about the future of powder so far as the architectural aluminum industry is concerned. I have seen some sample materials and am impressed with what the powder industry has done to become competitive in this respect, and I can state for a fact there are some excellent materials available today. Tomorrow there will be even more.

In Europe, the exterior extrusions and curtain walls of many buildings have been powder-coated for the past 20 years. I have seen many of these buildings and have assembled a good-sized collection of color photographs and slides of them. Buildings in the United States are now appearing with complete or partial powder-coated surfaces. Figure 3.6a shows one example, the World Savings Center in Oakland, California. Figure 3.6b shows the Terminal 2 building at the Port of Portland in Portland, Oregon.

Powder coatings are being used worldwide for the residential window market. The coatings available in the United States meet the AAMA 603.8 specifications. (AAMA are the initials for the American Architectural Manufacturers Association, a trade association of firms engaged in the manufacture and sale of architectural building components and related products.) These specifications for residential windows are available from AAMA. The coatings, if applied properly, provide an excellent exterior finish. The important point is to apply them *properly*. My favorite expression has always been, "To get a good finish, you need a good start." A good start—and a good finish—is the

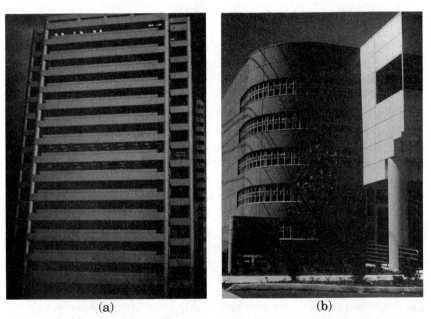

(a)   (b)

Figure 3.6   (a) The World Savings Center, Oakland, California, and (b) Terminal 2, the Port of Portland, Portland, Oregon. (*Both photos courtesy of Tiger Drylac Inc.*)

VOLUNTARY SPECIFICATION FOR HIGH PERFORMANCE ORGANIC COATINGS ON ARCHITECTURAL ALUMINUM EXTRUSIONS AND PANELS

1. SCOPE

1.1 This specification describes test procedures and requirements for high performance organic coatings applied to aluminum extrusions and panels for architectural products.

1.2 This specification covers factory-applied spray coatings only.

2. PURPOSE

The specification will assist the architect, owner and contractor to specify and obtain factory-applied organic coatings which will provide and maintain a superior level of performance in terms of film integrity, exterior weatherability and general appearance over a period of many years.

3. DEFINITIONS

3.1 The terms "film" and "coating" are used interchangeably in this specification and are defined as meaning the layer of organic material applied to the surface of the aluminum.

3.2 Exposed surfaces are those surfaces indicated by architectural drawings which are visible when the coated product is installed. These may include both closed and open positions of operating sash, ventilators, doors or panels.

4. GENERAL

4.1 To meet this specification, products tested shall meet all of the requirements of this specification.

4.2 Coatings shall be visibly free from flow lines, streaks, blisters or other surface imperfections in the dry-film state on exposed surfaces.

4.3 The total dry-film thickness on exposed surfaces, except inside corners and channels, shall be a minimum of 1.2 mils.

NOTE: When recommended by the manufacturer, film thickness specified may be increased to be consistent with the color selection and type of coating.

On two-coat applications involving a primer, the top coat shall be 1.0 mil minimum with a primer of 0.3 ± 0.1 mil.

4.4 Minor scratches and blemishes shall be repairable with the coating manufacturer's recommended product or system. Such repairs shall match the original finish for color and gloss and shall adhere to the original finish when tested as outlined in 7.4.1.1, Dry Adhesion. After application, allow the repair coating to dry for at least 72 hours at 65-80°F before conducting the film adhesion test.

NOTE: The size and number of touch-up repairs should be kept to a minimum.

5. TEST SPECIMENS

5.1 Test specimens shall consist of finished panels or extrusions representative of the production coated aluminum. A sufficient number of specimens on which to conduct instrument measurements with flat coated surfaces of at least 6" long and 3" wide, shall be submitted to the testing laboratory. The coating applicator or fabricator shall indicate exposed surfaces or submit drawings. Tests shall be performed on exposed areas as indicated on drawings or as marked on test specimens.

6. METAL PREPARATION AND PRE-TREATMENT

NOTE: A multi-stage cleaning and pre-treatment system is required to remove organic and inorganic surface soils, remove residual oxides, and apply a chemical conversion coating to which organic coatings will firmly adhere.

6.1 The products used to form the chemical conversion coating on aluminum extrusions and paneling shall conform with ASTM D 1730, Type B, Method 5 (Amorphous Chromium Phosphate Treatment) or Method 7 (Amorphous Chromate Treatment).

6.2 The coating weight of the chemical conversion coating shall be a minimum of 30 mg. per ft.^{2} on exposed surfaces as specified in ASTM B 449, Section 6, Class I. Processing shall conform with that specified in ASTM B 449, Section 5.

NOTE: Frequent in-plant testing and control of pretreatment is required to insure satisfactory performance of the coating system.

7. TESTS

7.1 COLOR UNIFORMITY

7.1.1 Procedure
Check random samples visually under a uniform light source such as a MacBeth daylight lamp or the North daylight sky. Samples must meet minimum dry-film thickness requirements.

7.1.2 Performance
Color uniformity shall be consistent with the established color range.

NOTE: Limits for acceptable production color variations are to be established between the approval source and applicator.

7.2 SPECULAR GLOSS

7.2.1 Procedure
Measure in accordance with the latest issue of ASTM D 523 using a 60° gloss meter. Samples must meet minimum dry film thickness requirements.

7.2.2 Performance
Gloss values shall be within ± 5 units of the manufacturer's specification.

NOTE: Standard gloss range reference values are:

GLOSS COLORS	SPECULAR GLOSS VALUE
High	80-Over
Medium	20-79
Low	19 or less

Figure 3.7   AAMA 605.2–90 specification sheets. (*Courtesy of the American Architectural Manufacturers Association*)

essential ingredient in preparing materials that are to be exposed to the ravages of weather on the exterior of a building.

The commercial specifications for the powder coating of aluminum extrusions are covered in AAMA 605.2–90. To give you an idea of spec-

7.3 DRY FILM HARDNESS

7.3.1 Procedure
Using a Berol Eagle Turquoise pencil, grade F minimum hardness, leave a full diameter of lead exposed to the length of 1/4" minimum to 3/8" maximum. Flatten the end of the lead 90 degrees to the pencil axis using fine-grit sand or emery paper. Hold the pencil at 45° to the film surface and push forward about 1/4" using as much downward pressure as can be applied without breaking the lead. Reference ASTM D 3363.

7.3.2 Performance
No rupture of film per ASTM D 3363.

7.4 FILM ADHESION

7.4.1 Procedure

7.4.1.1 Dry Adhesion
Make 11 parallel cuts, 1/16" apart through the film. Make 11 similar cuts at 90 degrees to and crossing the first 11 cuts. Apply tape (Permacel 99 or equivalent of sufficient size to cover the test area) over the cuts by pressing down firmly against the coating to eliminate voids and air pockets. Sharply pull the tape off at a right angle to the plane of the surface being tested. Test pieces should be at ambient temperature (approximately 65-80°F).

7.4.1.2 Wet Adhesion
Make cuts as outlined in 7.4.1.1. Immerse the sample in distilled or deionized water at 100°F for 24 hours. Remove and wipe the sample dry. Repeat the test specified in 7.4.1.1 within 5 minutes.

7.4.1.3 Boiling Water Adhesion
Make cuts as outlined in 7.4.1.1. Immerse the sample in boiling distilled or deionized water (210-212°F) for 20 minutes. The water shall remain boiling throughout the test. Remove the sample and wipe it dry. Repeat the test specified in 7.4.1.1 within 5 minutes.

7.4.2 Performance
No removal of film under the tape within or outside of the cross-hatched area or blistering anywhere on the wet test specimen. Report loss of adhesion as a percentage of squares affected, i.e., 10 squares lifted as 10% failure.

7.5 IMPACT RESISTANCE

7.5.1 Procedure
Using a 5/8" diameter round-nosed impact tester (160 in.-lb. range) such as a Gardner impact tester, apply a load directly to the coated surface of sufficient force to deform the test sample a minimum of 0.10". Apply tape (Permacel 99 or equivalent of sufficient size to cover the test area) by pressing down firmly against the coating to eliminate voids and air pockets. Sharply pull the tape off at a right angle to the plane of the surface being tested. The test specimen temperature should be 65-80°F.

7.5.2 Performance
No removal of film to substrate.

NOTE: Minute cracking at the perimeter of the concave area of the test panel is permissible but no coating pick-off should be apparent.

7.6 ABRASION RESISTANCE

7.6.1 Procedure
Using the falling sand test method, ASTM D 968, the Abrasion Coefficient shall be calculated according to the following formula:

ABRASION COEFFICIENT, LITERS PER MIL = V/T

WHERE:
V = VOLUME OF SAND USED IN LITERS
T = THICKNESS OF COATING IN MILS (0.001")

7.6.2 Performance
The Abrasion Coefficient Value of the coating shall be 20 minimum.

7.7 CHEMICAL RESISTANCE

7.7.1 Muriatic Acid Resistance (15 Minute Spot Test)

7.7.1.1 Procedure
Apply 10 drops of 10% (by volume) solution of muriatic acid (37% commercial grade hydrochloric acid) in tap water and cover it with a watch glass, convex side up. The acid solution and test shall be conducted at 65-80°F. After a 15 minute exposure, wash off with running tap water.

7.7.1.2 Performance
No blistering, and no visual change in appearance when examined by the unaided eye.

7.7.2 Mortar Resistance (24 Hour Pat Test)

7.7.2.1 Procedure
Prepare mortar by mixing 75 grams of building lime conforming to ASTM C 207 and 225 grams of dry sand, both passing through a 10-mesh wire screen with sufficient water, approximately 100 grams, to make a soft paste. Immediately apply wet pats of mortar about 2 square inches in area and 1/2" in thickness to coated aluminum specimens which have been aged at least 24 hours after coating. Immediately expose test sections for 24 hours to 100% relative humidity at 100°F.

7.7.2.2 Performance
Mortar shall dislodge easily from the painted surface, and any residue shall be removable with a damp cloth. Any lime residue should be easily removed with the 10% muriatic acid solution described in 7.7.1.1. There shall be no loss of film adhesion or visual change in appearance when examined by the unaided eye.

7.7.3 Nitric Acid Resistance

7.7.3.1 Procedure
Fill an eight-ounce wide-mouth bottle one-half full of nitric acid, 70% ACS reagent grade[1]. Place the test panel completely over the mouth of the bottle painted side down, for 30 minutes. Rinse the sample with tap water, wipe it dry, and measure any color change after a one-hour recovery period.

[1]The assay of the nitric acid (HNO_3) should be Fisher A-200 or equivalent; minimum 69.0%, maximum 71.0%.

Figure 3.7 *(Continued)* AAMA 605.2–90 specification sheets. (*Courtesy of the American Architectural Manufacturers Association*)

ifications for these finishes, which are used on exterior components of commercial buildings, I list the up-to-date specifications in Figure 3.7. The important properties of exterior coatings are, of course, long-range appearance and durability. One of the AAMA 605.2–90 requirements is a 5-year weathering test in Florida; 1990 is the fifth year of tests for some powder manufacturers. There is also included in the specifications a 3000-hour salt-spray resistance test and a 3000-hour humidity test.

7.7.3.2 Performance
Not more than 5 Δ E Units (Hunter) of color change, calculated in accordance with ASTM D 2244, when comparing measurements on the acid-exposed painted surface and the unexposed surface.

7.7.4 Detergent Resistance

7.7.4.1 Procedure
Prepare a 3% (by weight) solution of detergent and distilled water. Immerse at least 2 test specimens in the detergent solution at 100°F for 72 hours. Remove and wipe the samples dry. Immediately apply tape (Permacel 99 or equivalent 3/4" wide tape) by pressing down firmly against the coating to eliminate voids and air pockets. Place the tape longitudinally along the entire length of the test specimens. If blisters are visible, then the blistered area must be taped and rated. Sharply pull off at a right angle to the plane of the surface being tested. Detergent composition is as follows:

TECHNICAL GRADE REAGENTS	% BY WEIGHT
Tetrasodium Pyrophosphate	45
Sodium Sulphate, Anhydrous	23
Sodium Alkylarylsulfonate*	22
Sodium Metasilicate, Hydrated	8
Sodium Carbonate, Anhydrous	2
Total	100
*Allied Chemical Co. Nacconal 90F	

7.7.4.2 Performance
No loss of adhesion of the film to the metal. No blistering and no significant visual change in appearance when examined by the unaided eye.

7.8 CORROSION RESISTANCE

7.8.1 Humidity Resistance

7.8.1.1 Procedure
Expose the sample in a controlled heat-and-humidity cabinet for 3,000 hours at 100°F and 100% RH with the cabinet operated in accordance with ASTM D 2247.

7.8.1.2 Performance
Formation of blisters not to exceed "Few" blisters Size No. 8, as shown in Figure 4, ASTM D 714.

7.8.2 Salt Spray Resistance

7.8.2.1 Procedure
Score the film sufficiently deep to expose the base metal using a sharp knife or blade instrument. Expose the sample for 3,000 hours according to ASTM B 117 using a 5% salt solution. Remove and wipe the sample dry. Immediately apply tape (Permacel 99 or equivalent of sufficient size to cover the scored area) by pressing down firmly against the coating to eliminate voids and air pockets. Sharply pull the tape off at a right angle to the plane of the surface being tested.

7.8.2.2 Performance
Minimum rating of 7 on scribe or cut edges, and a minimum blister rating of 8 within the test specimen field, in accordance with the following Table 1 and Table 2 (Reference-modification of ASTM D 1654).

TABLE 1 — RATING OF SCRIBE FAILURE

MAXIMUM MEASUREMENT OF FAILURE FROM SCRIBE		RATING BY NUMBER
(in.)	mm	
0	0.0	10
1/64	0.4	9
1/32	0.8	8
1/16	1.6	7
1/8	3.2	6
3/16	4.8	5
1/4	6.4	4
3/8	9.5	3
1/2	12.7	2
5/8	15.9	1
1 or more	25.0 or more	0

TABLE 2 — RATING OF AREA OTHER THAN SCRIBE (Blisters, Corrosion, Etc.) (See Note.)

DESCRIPTION OF FAILURE (%)	RATING BY NUMBER
No failure	10
1	9
2	8
5	7
7 to 10	6
7 to 10 larger spots	5
11 to 25	4
26 to 40	3
41 to 60	2
61 to 75	1
Over 75	0

NOTE: The use of a ruled plastic grid is recommended as an aid in evaluating this type of failure. A 1/4" (6.4mm) grid is suggested as most practical for the usual specimen. In using the grid, the number of squares in which one or more points of failure are found is divided by the total number of squares covering the specimen. Convert this to a percentage figure.

7.9 WEATHERING

7.9.1 The coating shall maintain its film integrity and as a minimum meet the following color retention, chalk resistance, gloss retention and erosion resistance properties. The architect, owner, or contractor should request data relative to the long-term durability of the color(s) selected. Exposure panels must be made available to the architect or owner upon request.

7.9.1.1 Test Site and Duration
Test sites for on-fence testing are acceptable as follows: Florida exposure South of latitude 27° North at a 45° angle facing South for five years.

7.9.1.2 Color Retention

7.9.1.2.1 Performance
Maximum of 5 Δ E Units (Hunter) Color change as calculated in accordance with ASTM D 2244-85. Section 6.3 after the exposure test per 7.9.1.1. Color change shall be measured on the exposed painted surface which has been cleaned of external deposits with clear water and a soft cloth and corresponding values shall be measured on the original retained panel or the unexposed flap area of the panel. A portion of the exposure panel may be washed lightly

Figure 3.7 (*Continued*) AAMA 605.2–90 specification sheets. (*Courtesy of the American Architectural Manufacturers Association*)

The results of the AAMA 605 tests should give powder coatings an entry into the architectural aluminum market.

Architects should become increasingly aware of the potential of powder, particularly its durability.

It is somewhat ironic that because of the air pollution in the United States, particularly in southern California, it may soon be impossible to view the upper levels of our high-rise buildings. It should give arto remove surface dirt only. Heavy scrubbing or any polishing to remove chalk formation or restore the surface is not permitted where color measurements are made. New colors may be qualified without the exposure test per 7.9.1.1 provided that they are produced with the same pigments in the same coating resin system as a color on which acceptable five (5) year test data is available and which is within ± 10 Hunter Units in lightness (L).

7.9.1.3 Chalk Resistance

7.9.1.3.1 Performance
Chalking shall be no more than that represented by a No. 8 rating based on ASTM D 659 after test site exposure (per 7.9.1.1). Chalking shall be measured on an exposed, unwashed painted surface.

7.9.1.4 Gloss Retention

7.9.1.4.1 Procedure
Measure 60° gloss of exposed and unexposed areas of a test site exposure panel (per 7.9.1.1) as described in 7.9.1. Following ASTM D 523-85. The exposure panel may be washed lightly with clear water and a soft cloth to remove loose surface dirt. Heavy scrubbing or any polishing to restore the surface is not permitted where gloss measurements are made.

7.9.1.4.2 Performance
Gloss retention shall be greater than 30 percent after the exposure test per 7.9.1.1 expressed as:

$$\text{Percent Retention} = \frac{60° \text{ gloss exposed}}{60° \text{ gloss unexposed}} \times 100\%$$

7.9.1.5 Resistance to Erosion

7.9.1.5.1 Procedure
Measure dry film thickness of exposed and adjacent unexposed areas of exposure panels per 7.9.1.1, following procedures described in 7.9.1 using an eddy current meter as defined in ASTM B 244 or other instrumental methods of equal precision.

7.9.1.5.2 Performance
Less than 10 percent film loss after the exposure test per 7.9.1.1 expressed as a percent loss of total film:

$$\text{Percent Erosion} =$$
$$100\% - \left[\frac{\text{Dry film thickness exposed}}{\text{Dry film thickness unexposed}} \times 100\% \right]$$

7.10 Sealant Compatibility

The fabricator of the finished products should consult the sealant supplier in the selection of sealant which will exhibit adequate adhesion to the painted aluminum surface. Panel exhibits of the specific coating to be used should be submitted to the sealant manufacturer for tests and recommendations. It is recommended that the sealant manufacturer reconfirm tests with the actual materials used in production.

8. TEST REPORTS

8.1 The test reports on file with the applicator shall include the following information:

8.1.1 Date when tests were performed and date of issue of report.

8.1.2 Identification of organic coating and/or coating system tested, including production date, batch number, cure conditions and pre-treatment data, manufacturer's name and company submitting coated samples used in test.

8.1.3 Copy of drawings submitted showing exposed surfaces.

8.2 Test Results.

8.2.1 A statement indicating that the organic coating and/or coating system tested passed all tests or failed one or more.

8.2.2 In the case of a failure, which test(s) and a description of the failure(s).

8.2.3 Statement that all tests were conducted in accordance with this standard.

8.2.4 Name and address of the laboratory which conducted these tests and issued the report.

AAMA, 2700 river road, des plaines, illinois 60018

Figure 3.7 (*Continued*) AAMA 605.2–90 specification sheets. (*Courtesy of the American Architectural Manufacturers Association*)

chitects some comfort to know that properly applied powder coatings can reduce air pollution and thereby perhaps increase the visibility of their buildings.

Dealing with Rejected Parts

In any manufacturing process, there is always the possibility that parts, or even the product itself, will be rejected. If your product is rejected for reasons other than faulty finish, I can't help you. But if the problem is in the finish, let's consider the possible causes.

If there is a visible pretreatment failure, check your pretreatment system. If the system is titrated and working properly, the problem probably is due to a new type of contaminant on your substrate. This situation will require some help from your chemical technician.

If the problem is not in the pretreatment system, then your powder-coated reject amounts to either a powder problem or an application problem. The powder problem may require the help of your powder technician. The powder-application problem you can deal with yourself. If it is a touch-up problem when the part is first being coated, the people doing the touch-up, once alerted, can usually solve it.

If the product can be repaired, how do we repair it? Some "light," or lightly coated, products can be repaired with a light recoating of powder and a rebaking of the product. If the damage to the finished coating is severe, there are two alternatives. Lightly sand the affected areas and recoat, or strip the part and start over.

If undercure is the problem, it's usually possible to put the piece through the oven for a second time, in order to effect a complete cure. But before you decide to do that, you should investigate the ability of your powder to undergo a second trip into the oven (earlier in this chapter we discussed the overbaking capabilities of various powders). One way to find out is to try. If you can eliminate the problem with a second trip through the oven, you won't need a new type of powder.

If the cosmetic effect is extremely important in your finish, check the part under a good lighting system prior to putting it in the oven. It is much easier to recoat, or to blow the uncured powder off the part prior to the cure cycle, than to strip it later.

If your product has been shipped and arrives at its destination with a damaged finish, most powder suppliers can help you obtain spray cans of matched wet, air-dry coatings to repair the blemished or lightly damaged surface.

4

Pretreatment

Overview

Before we get into the analysis and breakdown of a pretreatment system, let's look at what we're trying to do to the surface of the part. Basically, we're trying to prepare it for a coating of powder, which is dry and clean. During the curing process, the powder will need to "grab" or "cling to" the surface of the part. For this to happen, the surface must be

1. *Totally clean:* All soils, both organic and inorganic, must have been removed.

2. *Totally dry:* After its cleaning and pretreatment, the part must be completely dried by an oven with a large amount of hot air.

3. *Totally enhanced:* The surface of your part must be elevated to a higher degree of preparedness. It must be put into a slightly acidic condition during its pretreatment; chances are it will have received treatment first with an alkaline material (to remove organic soils) and then a chemical of an acidic material.

If you've had any experience with a finishing system, you've heard of those three requirements in relation to cleaning or pretreatment. And you probably didn't pay much attention to them. But now, it's essential that you learn what it means to prepare your parts properly before they're powder-coated. Powder coatings are *dry* materials. They are applied dry. If there is a wet, or even a moist, contaminant on the surface when the newly powder-coated part enters the oven, the moist contaminant will boil. The now-heated wet powder will mix and possibly boil, and a barrier will be created between the substrate and the powder. When the cure is completed, there will be no bond at the point of contamination. So it is absolutely essential for your part to be totally *clean, dry,* and *enhanced.*

To be totally dry, your part must go through a dry-off oven, which will dry the liquids applied during the pretreatment process. Be aware that the oven will *not* dry out large puddles of water which have accumulated because of product design and cannot drain by gravity.

Contrary to what some people think, it takes more than a greasy rag to prepare your parts properly for powder coating, or for that matter, for any type of finish. Wet coatings and various plating processes also require pretreatment before the organic finish is placed on the parts. Be assured that pretreatment is not something that originated with powder coating; it's been around for a long time.

It's true that, in the past, enough attention was not always given to pretreatment prior to coating. But the manufacturing process has become very competitive, and manufacturers are demanding longer-lasting finishes for their products. In order to get this durability, you have to provide the best pretreatment available. Remember this, "To get a good finish, you need a good start."

Various Methods of Pretreatment

Pretreatment can be accomplished in many different ways. For example, you can use

1. The old rag-and-a-pail-of-solvent method
2. Sophisticated solvent systems
3. Abrasive blasting or cleaning
4. Aqueous chemicals

A rag and a pail of solvent. This method does nothing but move the grease and oil from one part of the surface to another. Thus the powder coating that is applied and eventually heated in the oven would be undergoing the cure while the greases and oils beneath it were boiling and bubbling. This, of course, would preclude a good bond.

Sophisticated solvent system. These systems can do a job cleaning the surface of your parts; but remember that the surface of the part must also be able to receive, and to bond with, the cured powder. Pretreatment systems are designed to clean your substrate *and* to enhance its surface with an acidic surface so that the powder coating will adhere properly.

Abrasive blasting or cleaning. In this system, compressed air is sometimes used as a method of propelling the blasting media. A centrifugal wheel is also sometimes used. The wheel, with the media placed within it, rotates and by centrifugal force propels the media onto the substrate with a great deal of force, blasting it clean. This wheel method is best suited for removing heat scale from hot-rolled steel, removing rust, and removing carbonized heat scale from welding.

There are times when the abrasive systems can help you by prepretreating the surface of hot-rolled steel, in order to give you a better surface upon which to place a phosphate material. The loose surface material and scale associated with hot-rolled steel will be removed to a great extent, and the part to be finished will have received an excellent pretreatment when the blasting is followed by cleaning, phosphate coating, and final rinse.

Some companies use one of the above three methods for pretreating their parts. But I suppose the majority of companies, those with high-speed production lines, use the fourth process, the aqueous cleaning system, and I will limit our discussion now to this method of pretreatment.

Aqueous chemicals. Aqueous chemicals can be applied using one of three tools: (1) a spray wand, (2) an immersion tank, and (3) a power washer.

1. *Spray-wand methods* are getting more sophisticated every day. Cleaning solutions, phosphate coating materials, and, yes, even sealers can be applied using wands. Spray wands can be used in conjunction with power-washer systems. If your part has portions partially hidden from the spray nozzles of a power washer, it's possible to prepare its surface prior to the time it enters the power washer, thus ensuring that the hidden locations will be pretreated and have the same longevity as the outer surface of the part. Spray-wand cleaning is usually done prior to using a power washer. Figures 4.1 and 4.2 show the external and internal views of a spray-wand system.

2. *Immersion-tank systems* are used for many reasons. Smaller parts which have areas hidden from the power-washer nozzles are sometimes handled this way. Sometimes aluminum extrusions, because of the actual extrusion profile or shape, are preloaded on hanging fixtures, pretreated in immersion tanks, dried, then coated and cured on a normal conveyor line.

3. Several years ago, I felt a client of ours should place his new finishing system on hold, until his company reevaluated their pretreatment plans. They had decided that an immersion system was really what they wanted. We suggested that they purchase a power washer rather than an immersion tank. Their judgment was based solely on the "economy" associated with the purchase price of a tank line versus

Figure 4.1   External view of a spray-wand system.   (*Courtesy of Fremont Industries, Inc.*)

that of an in-line power washer. I put my case to the client in the form of a list of questions. These questions in themselves will make the case for a *power-washer system.*

- Will the same dirty hands, wearing dirty gloves, which place dirty parts in an immersion-cleaning-system basket, be the hands that remove the clean pretreated parts?
- How will you remove fingerprints left on the clean surface by an employee transferring the part from the immersion-tank baskets to the overhead-conveyor-system hanging fixtures?
- How do the cleaning solution, phosphate material, seal rinse, and plain-water rinses reach those parts that are leaning against each other and thus preventing a turbulent-tank-line system from doing its job?
- When you remove your parts basket from the various stages, what happens to the parts at the bottom of the basket as the chemicals from above drain by gravity? Are you certain that all the material will properly drain before the part goes into the next stage? Are you certain there will not be contamination by carryover?
- How do you propose to dry the parts when they have completed the pretreatment cycle? Will you first remove them from the baskets,

hang them on hanging fixtures, and put them in a dry-off oven? (They'll probably dry in the ambient air, start to oxidize, and flash-rust.) Or will you put the entire basket into a batch-type dry-off oven and hope that all of the liquid puddles evaporate? (Any chemical engineer will tell you the futility of hoping that liquid puddles will evaporate.)

The real loser in my book is the company that spends much of its time trying to build a quality product—maybe even the best product in its field—and then, in an effort to cut costs, skimps in the finishing cycle by using a poor pretreatment system. When designing your finishing system, ask yourself some pointed questions: If this part gets sold and shipped to Podunk, USA, is there enough profit in it to

1. Pay for shipping back to us if the finish falls off in transit?
2. Ship another part to the angry customer or distributor if the finish falls off?
3. Strip and/or disassemble the part, refinish it, reassemble it, and reship it?
4. Use a spot on the conveyor system where a new part could be hanging to hang the part to be refinished?

It looks to me as if we are talking about, at the very least, a four-time finishing cost.

Pretreatment, Phosphatizing and/or Conversion Coatings

When you get around to writing your first book, you'll find yourself making many changes each time you rewrite a draft. That's exactly what has happened to me in working on this chapter. I wanted to call the chapter "Phosphatizing," a term many of us in the industry are familiar with. But I also wanted to call it "Pretreatment," which is what I finally settled on. The main reason for the dilemma in deciding on a title is that many people pretreat steel. Steel gets phosphatized. Other materials, like aluminum, get a conversion coating—the surface composition of the substrate is converted. Actually, the steel surface gets converted too, if you think about it. And there is the special preparation of brass, plated brass, bronze, and other plated materials. We enhance all these surfaces.

I'm not trying to confuse you with literary or chemical semantics. Please read on, and things will become clear to you, I promise. If you read in its entirety no other chapter in this book, please read all of this

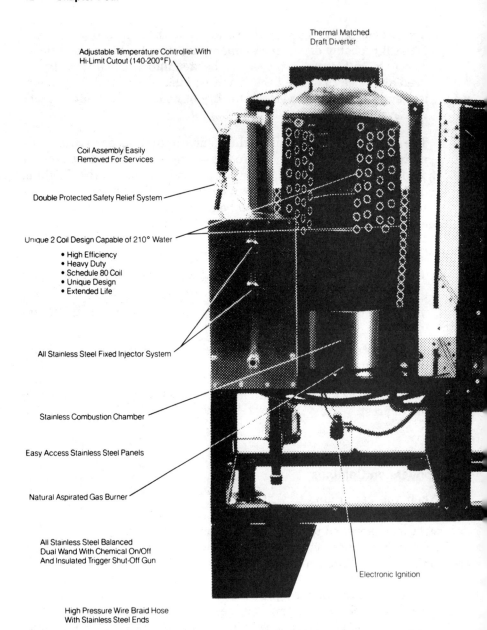

Thermal Matched
Draft Diverter

Adjustable Temperature Controller With
Hi-Limit Cutout (140-200°F)

Coil Assembly Easily
Removed For Services

Double Protected Safety Relief System

Unique 2 Coil Design Capable of 210° Water

- High Efficiency
- Heavy Duty
- Schedule 80 Coil
- Unique Design
- Extended Life

All Stainless Steel Fixed Injector System

Stainless Combustion Chamber

Easy Access Stainless Steel Panels

Natural Aspirated Gas Burner

All Stainless Steel Balanced
Dual Wand With Chemical On/Off
And Insulated Trigger Shut-Off Gun

Electronic Ignition

High Pressure Wire Braid Hose
With Stainless Steel Ends

Figure 4.2   Internal view of a spray-wand system.   (*Courtesy of Fremont Industries, Inc.*)

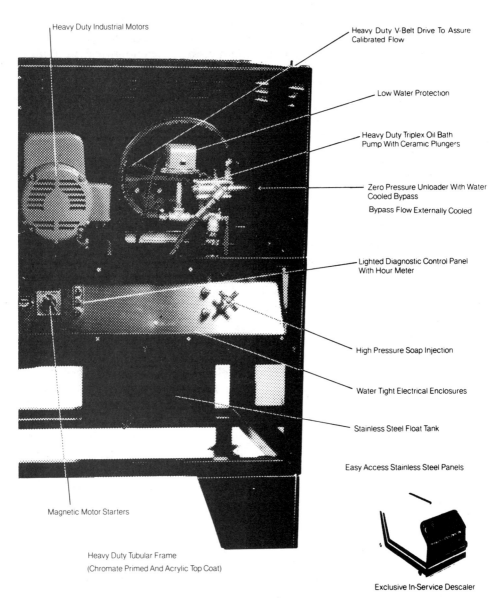

Heavy Duty Industrial Motors

Heavy Duty V-Belt Drive To Assure Calibrated Flow

Low Water Protection

Heavy Duty Triplex Oil Bath Pump With Ceramic Plungers

Zero Pressure Unloader With Water Cooled Bypass

Bypass Flow Externally Cooled

Lighted Diagnostic Control Panel With Hour Meter

High Pressure Soap Injection

Water Tight Electrical Enclosures

Stainless Steel Float Tank

Easy Access Stainless Steel Panels

Magnetic Motor Starters

Heavy Duty Tubular Frame
(Chromate Primed And Acrylic Top Coat)

Exclusive In-Service Descaler

Figure 4.2 *(Continued)* *(Courtesy of Fremont Industries, Inc.)*

one. Then read it again, and again, until you completely understand pretreatment. Then read it one more time.

During my tenure in the finishing industry, I have come to realize more and more just how important a good pretreatment system is for a long-lasting finish on any painted product. When powder manufacturers talk about the superiority of their very best grades of powder, they know that, without pretreatment, these miracles cannot be produced. (Yes, I say "miracles," because even in their brief life to date powder coatings have performed miracles.) The salt-spray resistance tests, impact-resistance tests, and other such tests all give their very best results with a good pretreatment.

Lehr's First Law: **To get a good finish, you need a good start.**

Some Cautionary Notes

If you have been looking at powder samples—either those samples coated on panels or on your own parts—don't embarrass yourself or the powder salesperson by giving the coating the old test of scratching the surface with a nickel. Many times samples are made merely to show the potential customer what the color looks like; they are made on clean, but not pretreated, panels.

Let me digress for a moment here. I can remember attending a corporate meeting at which a powder sales engineer displayed some color panels to demonstrate a particular color match. I will never forget the argument which started when, after hearing of all the excellent virtues of powder, the client peeled the cured coating from the surface of the thin panel presented to him by the sales engineer. Then the client really got into an attack mode, quickly peeling powder from a second panel, and a third and a fourth and finally the fifth panel, on which he scratched a simulated crosshatch and commented on how it could now serve for a tic-tac-toe game. The poor sales engineer and I were finally able to get the client's attention and explain to him what the panels were for, and how they were made. I think he eventually made his decision to buy only on the basis of the embarrassed position he found himself in when he realized that the panels were made for display, not for testing. I am certain the powder would have passed all of the tests the sales engineer cited, but only if applied to a properly pretreated panel—properly prepared the same way your parts will be prepared on your finishing line.

One more cautionary note before we continue: To have a good pretreatment system, you must have water of a fairly decent quality. Before you get started, have your potential or present chemical supplier check the quality of the water you will be using on the finishing line.

After all, it's the same water with which you'll be mixing your supplier's chemicals. If the water quality is poor, your pretreatment will also be poor. You may find it necessary to pretreat your water before using it. Later in this chapter we will discuss ways to remedy the situation when the quality of your water is not up to standard.

Pretreatment of Steel

Steel is probably the material most commonly coated with powder, so we will start out discussing its pretreatment. Believe it or not, much of the steel used in the United States is imported. The next steel item you buy could conceivably be made from truly international steel. Its different parts could be made from steels manufactured in Europe, Japan, and the United States. Your purchasing department is always trying to get the best value for your company. Today the best value could be steel made in Germany; next month, steel made here at home or in the Far East.

One of the differences in steel is the coating placed on its surface to prevent it from oxidizing or rusting while it is being processed and then finished on your finishing system. This coating is usually a rust inhibitor or an oil. After the steel gets into your plant, your people cut it, bend it, form it, punch it, and place all types of cutting oils on it. Some of these contain waxes and corrosive additives. Can you imagine the result of placing wax and cutting oils together, then placing them along a seam that will eventually be welded? Baked contaminants! All of which must be removed during pretreatment so that some family can eventually beat the poor product to death, but still expect the finish to last and the product to look good.

The five-stage iron phosphate cleaning system

Figure 4.3 illustrates a typical five-stage power-washer system.

Stage 1. As we have noted, oils, greases, and waxes (both baked if your parts are welded and fresh if deposited during the manufacturing process) will be found on the surface of your steel. The first stage of your power washer must eliminate all of these contaminants if your pretreatment process is to work properly. Not only that, the washer must do it within a specified period of time. You should plan on at least 1 1/2 to 2 minutes for this first stage. Your chemical supplier can tell you the exact time by running tests. Think, 2 minutes! If management or your chemical supplier says you can clean in less than 2 minutes—whether because you are budget-conscious or because of the

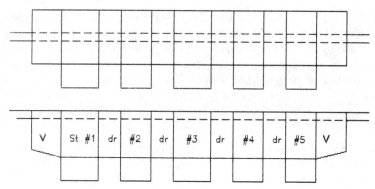

Figure 4.3 Typical five-stage power-washer system. Stage 1: alkaline cleaner; stage 2: ambient rinse; stage 3: iron phosphate coating; stage 4: ambient rinse; stage 5: seal rinse. Entrance and exit vestibules (V) are located at each end of the washer.

superior parts you are buying—ask these people who will take the responsibility when, a year from now, you change your steel supplier and find that the parts now need more time in the washer to remove a cosmoline-type material. Will your management, chemical supplier, or subcontractor "buy" the extra few feet you'll need in your power washer to do the job? I doubt it very much. There's usually no place at this point to hang your hat. You designed, you specified, you take the blame.

We were recently involved in a situation like that. We had designed a system for a client company and had run enough tests between the time of the design and the installation of the system to satisfy everyone, including ourselves. But the manufacturing division of the company had in the meantime added a new process which created a waxy film on the surface of the part to be pretreated. The original pretreat solution, an alkaline cleaning agent, was unable to completely remove the waxy film. The chemical supplier, seeing the problem, went to a customer's nearby plant and borrowed sufficient quantities of a booster chemical to remove the waxy material. But it took all of the 2 minutes allocated to the first stage to do the job. Just remember that 2 minutes is not too long for parts that have accumulated contaminants since the steel was manufactured.

You will notice that I mentioned oils, greases, and waxes but not rust, or oxidation, among the contaminants. The alkaline cleaners used in stage 1 are not made to remove rust. If you have rust problems, discuss it with your chemical supplier, and he or she can advise you what steps to take. Maybe extra cleaning stages will be needed to give you an acid cleaner and a second rinse. Of course, the proper thing to do is to eliminate the cause of the rust. This is the best and cheapest way to handle the problem.

Alkaline cleaners come in powder and liquid forms. Boosters and other additives to make the formula best-suited for your particular situation are added to the alkaline cleaner when it is being placed in the first stage of your washer. These boosters, additives, whatever, may not be necessary to clean your parts today, but will be available if needed in the future.

Alkaline cleaners have a good life span if they are set up properly. I have seen cleaning solutions that were inky black do a very good job. A good chemical representative will help you keep track of your system and will advise you when and how often you should drain the solution and replace it. The chemical representative will also tell you exactly how to dispose of the solutions when it comes time to change them.

Remember that a good chemical technician can help you cope with all of your cleaning problems. If you have a good technician, you have someone you can rely on.

In most plants where production is heavy, one employee is responsible for the frequent checking of chemical concentrations, pressures, temperatures, pH, if necessary, and all of the other things necessary to make certain the power washer is working properly at all times. If you like, programmable controllers can check the chemical solutions for you. The cost for this equipment is not cheap, but what is good production and peace of mind worth? There are many types of chemical replacement systems available.

Stage 2.   After the alkaline cleaner has done its work on the part, you will have to remove any excess alkaline cleaner, residue, or contaminated materials from the surface of the now-clean substrate before the phosphate coating is applied. For stage 2, we use an ambient plain-water rinse, of which a constant small percentage is overflowed from the tank to the sewer system. The idea is to keep the water as clean as possible. It would be dumb to replace contaminants on the clean surface you have just created, so keep the rinse overflowing. Usually 30 seconds is sufficient to do the job. However, if your part is extremely complex in shape, it may take a little longer. Discuss this with your chemical supplier and with the company that will build your washer.

A good overflow rinse will overflow contaminated surface water at the rate of about 10 percent per hour. It is prudent to drain your rinse tanks every night. If the residue in the bottom of the tanks is heavy, rinse the tanks out. Keep the rinse tanks as clean as possible. Measure the pH and the conductivity to determine your water quality. If the water itself is contaminated at the end of the day, change it.

Stage 3. The next stage of your washer works somewhat of a miracle, and a fascinating one. This is where the surface of the steel is coated with iron phosphate. I specify iron phosphate because, although there are other phosphate materials, they require more elaborate application systems.

Iron phosphate is applied at the rate of from 20 to 65 milligrams per square foot of surface. The coating thickness can be measured. But if it is applied properly and the concentration is checked frequently, it is not necessary to keep a constant check. Your chemical representative can help you choose the best method for checking the coating, and for keeping the system properly adjusted.

Proper concentrations, proper temperature, proper impingement, proper pH, and time are the necessary factors in applying iron phosphate coatings. I feel that a minute is a good time, and most chemical people will agree with me on this.

Iron phosphate coatings are usually derived from solutions which contain very little iron. Radiometric studies indicate that so-called iron phosphates consist of iron phosphate and gamma iron oxide. Whatever their composition, they are produced on ferrous metals through the combined use of acid phosphate salts and accelerators.

The interesting thing about this third stage is what actually happens to the steel surface when it is coated with this water-soluble liquid. The amorphous coating that now covers your steel, and the cured powder, will protect the surface of your parts for many years. This amorphous coating is the necessary basis for a rust-inhibiting surface for your substrate or part.

Stage 4. This good start must have a good finish, so let's continue to follow our part on its trip through the washer. The phosphate coating growth is not uniform in size and shape. Some crevices, much like the open spaces between your teeth, will remain on the surface. These crevices must be sealed and protected. Another ambient water rinse for about 30 seconds will stop the action of phosphate and prepare the part for the next stage. This ambient rinse should also be overflowed, and the tank drained daily, as was done in stage 2.

Stage 5. Now we enter the final stage of our pretreatment system— the final rinse, seal rinse, compounded chromic acid rinse, acidic rinse…. It's known by many names within the industry. In this stage, those spaces between the phosphate crystal growth must be filled in to completely cover and seal the now-passivated metal surface or the cleaned surface will immediately begin to oxidize, or rust.

Compounded chromic acid rinses were popular for years, but for many reasons, the industry seems to have gone to "nonchromic" rinses. Even so, there is still much discussion as to how much protection a chromic acid rinse will give. Most people, especially the conservative ones, will tell you that it will give at least 2 times as much salt-spray resistance. Oh, for those days of chromic acid rinses!

The waste disposal of chromic acid is a problem, but it can be handled with special equipment recently developed by the chemical industry. Thanks to many thinking people, there are now methods that allow you to use compounded chromic acid rinses on your finishing line, so you can enjoy the excellent protection they give the surface of your product. Figure 4.4 shows the Recycle Seal system, which incorporates one of these methods. Other systems include a filter press.

The final position of your product in the marketplace is the most important factor in deciding whether you need a compounded chromic acid rinse or not. This rinse is, I feel, like the fudge topping, whipped cream, and cherry on top of plain vanilla ice cream.

Lehr's Second Law: **Don't pretreat today what you can't powder-coat and cure today.**

Your chemically pretreated parts need the assistance of the cured powder to make them function well, and the powder needs the assistance of the chemical pretreatment to help it adhere to the surface of your metal. It's a two-way street; you pretreat, you must coat and cure. If not, the pretreated surface will start to oxidize and rust; it will also pick up airborne oils and dust.

Figure 4.4 The Recycle Seal, which incorporates a chromic acid rinse. (*Courtesy of Fremont Industries, Inc.*)

The three-stage iron phosphate system

We have just described the five-stage power-washer setup for an iron phosphate bath system. How does this system differ from a three-stage line? We can perhaps compare the two succinctly before looking in depth at the three-stage system. See Table 4.1.

See the difference? In the three-stage washer, it is necessary to use the cleaner and the phosphate together in the first stage. Because the phosphate solution is an acid, the cleaner, which would normally be an alkaline substance, must be replaced with an acidic material to complement the acidity of the phosphate. Acidic cleaners are not the best materials for removing standard shop oils and greases, and they won't always get the job done.

Another point to consider: In a three-stage system, it is always necessary to dump the entire stage when the cleaner needs replacing. Not so with a five-stage system, where you can get extra life out of the chemicals, since stage 1 or stage 3 can be dumped individually as needed.

Most chemical people will agree (even those who sell the three-stage system chemicals) that a five-stage system does the better job. It also gives you the advantage of being able to install additives in the cleaning solution should the need arise for extra cleaning help. Figure 4.5 shows a typical three-stage iron phosphate power-washer system.

Other phosphate systems

Zinc phosphate coatings have excellent exterior durability, but their application is more complex than that of iron phosphate coatings. It usually requires at least one more stage to further enhance the substrate before the application of the zinc phosphate. The zinc phosphate is also applied as a much heavier coating than iron phosphate; it is applied in a range of from 100 to 300 milligrams per square foot, compared with the range for iron phosphate of from 20 to 65 milligrams. More sophistication is required in the power washer itself for handling the sludge material created by the zinc phosphate. The pretreatment

TABLE 4.1   Comparison of Three-Stage and Five-Stage Iron Phosphate Washers

	Three-stage	Five-stage
Stage 1	Acidic cleaner mixed with phosphate solution, 3–5 pH	Alkaline cleaner only, 10–12 pH
Stage 2	Ambient rinse	Ambient rinse
Stage 3	Final rinse	Phosphate coating, 2.5–5.5 pH
Stage 4	None	Ambient rinse
Stage 5	None	Final rinse, 2.5–5 pH

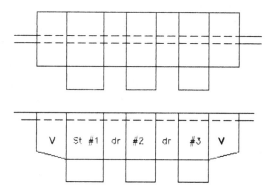

Figure 4.5 Typical three-stage power-washer system. Stage 1: acidic cleaner/iron phosphate combination solution; stage 2: ambient rinse; stage 3: seal rinse. Entrance and exit vestibules (V) are located at each end of the washer.

system must be controlled by a responsible person, and titration testing and other maintenance practices must be performed at regular times.

There are other aqueous phosphate systems used on steel, but none I know of are used in conjunction with powder coating systems.

Phosphate immersion-tank systems

In most cases, immersion-tank systems work with a chain of chemicals similar to those used in power-washer systems. Sometimes stronger additives are used to help remove stubborn soils. Remember, a tank system does not have the same impingement of chemicals onto the surface of the substrate as is available in power washers. Recirculation of chemicals by pumps is all the force available in an immersion-tank system.

Pretreating hot-rolled steel

If you are using hot-rolled steel for your end product, it may take more than a simple five-stage cleaning system to do the job for you. Tests will have to be performed. You may even find it necessary to pretreat with abrasive blasting prior to the application of a phosphate coating material.

Pretreating welded parts

If there is any welding within your manufacturing process, you must be sure to choose the pretreating method that pretreats the welded area completely. This is particularly important with overlapping welded seams. Figure 4.6 shows how oil is sometimes trapped in these seams. If, after pretreatment and dry off, the parts are powder-coated and sent to the cure oven, the oil trapped between the two welds starts to creep out, and the assembly is rejected. Some pretreatment methods, such as an alkaline bath prior to welding, can prevent this problem of oil bleedout. They also reduce smoke and residue as the welding process is completed.

Pretreatment of Aluminum

Some companies make products using an aluminum substrate in sheets, castings, or extrusions. The treatment of aluminum is usually different from that of steel. If aluminum is the substrate to be pretreated in your product, there are several basic types of systems that can be used:

1. Chrome oxide (amorphous chrome) treatment
2. Chrome phosphate treatment
3. Surface activation
4. Other systems

The *chrome oxide* system uses a hexavalent chrome, which makes the

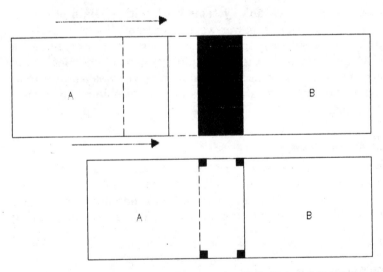

Figure 4.6 Oil trapped in a seam between A and B. If A and B are then welded at four spots, the part pretreated, powder-coated, and cured, the oil will start to expand and creep out under and between the welded spots with the heat of the cure oven. This contamination will mix with the coating and cause a reject.

surface of the substrate extremely corrosion-resistant, while the *chrome phosphate* system contains trivalent chrome, which is less corrosion-resistant than hexavalent chrome. Trivalent chrome phosphate coatings are generally more, shall we say, "mechanically" durable than the chrome oxide materials.

Table 4.2 shows a comparison of the chrome oxide and chrome phosphate systems. In both systems, the first stage usually provides some form of an alkaline cleaner to remove the organic soils; the second stage involves an overflowing ambient water rinse. In stage 3, an acidic material called a deoxidizer is used. It etches the surface and removes scale, smut, and other foreign materials. But it has another function—to "activate" the surface of the substrate in preparation for its trip into the chrome oxide or chrome phosphate stage. In other words, this deoxidizing enables the chrome treatment to become more functional. The fourth stage involves an ambient rinse; in the fifth stage, the chrome oxide or chrome phosphate is applied, and in the final stage in the immersion-tank system, the ambient rinse is repeated.

In the chrome phosphate system the seal rinse contains hexavalent chrome, which adds to the corrosion protection. The seal rinse is followed by a deionized (D/I) water rinse.

The third pretreatment method for aluminum products, *surface activation,* is a combination system that is used primarily when other materials such as steel might be pretreated on the same system. A three-stage system can be used with surface activation. The best you could ask for is a system for the cleaning of aluminum and steel that involves three stages: a cleaner/phosphate, an ambient rinse, and a fi-

TABLE 4.2   Chrome Oxide and Chrome Phosphate Systems for Pretreating Aluminum

Stage	Chrome oxide	Chrome phosphate
	With power washer	
1	Clean	Clean
2	Rinse	Rinse
3	Chrome oxide	Chrome phosphate
4	Rinse	Seal rinse
5		D/I rinse
	With immersion tank	
1	Clean	Clean
2	Rinse	Rinse
3	Deoxidize	Chrome phosphate
4	Rinse	Seal rinse
5	Chrome oxide	D/I rinse
6	Rinse	

nal rinse. This system could be made more elaborate by going to a five-stage system, which would provide a mediocre pretreat of multi-metals.

The fourth method of pretreating aluminum parts comprises a variety of systems, some of which are patented. They do not use chrome, and they are usually made up of a cleaner, a rinse, and an application of a proprietary material to prevent oxidation.

Pretreatment of Brass, Bronze, and Plated Materials

The main reason for using a pretreat system with brass, bronze, and plated materials would be to clear-coat the surface, in order to prevent oxidation. Your brass might even acquire a black nickel, relieved-type coating for an "antique" effect.

The pretreatment of these materials is actually a simple system, but it involves much preparation and extensive testing. There are many variables to be concerned with: Is the buffing compound used to polish the brass wet, soft, dry, or hard? Is it Tripoli or rouge? Is the surface coated with a relieved black nickel? Is it a plated or solid brass surface?

You must systematically answer all of such questions and then be able to properly clean, dry, coat, and cure the parts without spotting them. It can be done, and the process is simple, but it will vary with the condition of the part you offer the cleaning system.

My personal experiences have been with the use of an ultrasonic cleaning system using an alkaline cleaner, such as buffing compound, to remove any accumulated organic soils, followed by multiple fresh-water rinses, followed by a compounded chromic acid rinse, followed by a D/I rinse.

Pretreatment of Other Metals

There are other metals you may want to prepare for powder coating. Iron phosphate coatings can be used for zinc die-cast parts. Aluminum pretreat systems work well with aluminum die-cast and zinc sheet.

Various Factors in a Pretreatment System

pH

I do believe in giving credit where credit is due. As often as I have come across a definition of pH, I always walk away talking to myself. We know that a knowledge of pH is important in working on a pretreatment line; after all, we have to keep the pH in balance in our pretreatment system. But, try as I might, I always had trouble explaining the pH factor itself to people. Then one day I had a discussion with Brad Gruss of Fremont Industries,* and it was he who finally gave me a reasonable explanation for pH, one he got out of *Pressure Points,* a magazine devoted to the pressure cleaning industry.

Brad's explanation went like this:

> pH is the symbol for the logarithm of the reciprocal of hydrogen-ion concentration in gram-atoms per liter. For example, a pH of 5 indicates a concentration of 0.00001, or 10^{-5}, gram-atoms of hydrogen atoms in 1 liter of solution. The H stands for the concentration of the hydrogen atoms in the aqueous solution. In water, you have as many hydrogen atoms (positive) as you have hydroxyl ions (negative); therefore, water is neutral. We can write $H^+ = OH^-$, where OH^- = hydroxyl ion and H = / – OH = $H^2 O$ = water.

> So when we have more H^+ (hydrogen ions) than OH^- (hydroxyl ions) in an aqueous solution, then we are dealing with an acid. When we produce more negative hydroxyl ions, then we have an alkaline solution.

> The pH scale starts at 0, a very strong acid, and goes to 14, the strongest alkaline. At 7 we have our neutral point (water).

> Back in the "olden days," if you had watched your grandma stirring up a batch of laundry soap in the shed on the farm and you had asked her about pH, she would most likely have thought you were getting vulgar and given you a rap on your knuckles with the big wooden spoon she was using as a stirrer. And, if some of the hot brown soap had then dripped on your bare skin, you'd have realized her pH needed some adjusting. Back in those days, many a housewife made her own laundry soap using leftover fat and suet which she would collect until the smell drove her to dump it into a great kettle; boil it with lye until it had cooked into layers of liquid soap, glycerin, and brine; and then separate it into big cast slabs of brown soap. She would then cut these slabs into pieces to use for washing clothes. The soap was an ugly brown, and it could attack hands as well as dirt with a vengeance. Part of the powerful cleaning action came from the high alkalinity of the soap. In fact, it usually contained a certain amount of free alkali left over from unabsorbed lye. Grandma's lye topped the pH scale at 14.

> The standard dishwashing compound averages a pH of about 10, a mild buffered alkaline cleaner for aluminum is about the same, and strong alkaline cleaners for ferrous surfaces (iron or steel) range from 11 to 13. We can make use of the pH factor in assessing hand cleaners too. The sweat on our skin is mildly acidic; this means that hand cleaners should be at least at neutral (7) or mildly acidic (6). Tests have shown the skin can handle acid pHs down to 4.5 better than the alkali pHs of 8.5, because of the neutralization action of the alkali.

---

*When Brad was a baby, his father, Fremont Gruss, checked the pH of the contents of the bottle every time he fed Brad, claiming that the proper pH balance would give Brad a good disposition. As a result, Brad understands pH almost intuitively.

Alkali + acid (in equal amounts) = neutral salts = neutralization

You can see how pH numbers give us a convenient measuring scale. Because the scale is based on logarithms, the jumps are drastic, as revealed in the following table, which gives mathematical pH relationships:

pH number		Relative acidity or alkalinity
0	Acidity	10,000,000
1		1,000,000
2		100,000
3		10,000
4		1,000
5		100
6		10
7		1
8		10
9		100
10		1,000
11		10,000
12		100,000
13		1,000,000
14	Alkalinity	10,000,000

This means that if pure water is considered to be neutral, midpoint, the extreme of acidity (pH 0) would be 10 million times more acid than pure water, and extreme alkalinity (pH 14) would be 10 million times more alkaline than pure water.

What does this mean to us? To get a better handle on pH, let's examine another table, one which relates to items we are all familiar with. These figures obviously are averages:

Common materials	pH
Battery acid	0.1
Lemon juice	2.0
Ginger ale	3.0
Grapefruit	3.3
Bananas	4.5
Cow's milk	6.5
Pure water	7.0
Detergents	3.0–11.5
Sodium bicarbonate	8.4
Calcium carbonate	9.4
Laundry soap	10.0
Ammonia	11.0
Trisodium phosphate	12.0
Sodium hydroxide (lye)	14.0

Compare this table of common materials with the chart of pH relationships above and notice the huge range.

The pH is a very important standard in the finishing industry because it enables us to check cleaning, etching, and other chemical solutions in water. It is interesting to note that whether you dissolve 1 ounce or 10 ounces of alkaline cleaner in a gallon of water, you will get the same pH reading when you check it with your litmus paper. This is the reason we cannot totally rely on the pH reading with respect to the cleaning power or level in a given cleaning tank; we must rely instead on the concentration of ounces per gallon of chemical by titration.

Ambient rinses

Ambient rinses are an important part of your pretreatment system. They are normally placed between the chemical stages and are used both to stop a chemical surface activation and to prevent carryover from one stage of a chemical system into another, that is, prevent the alkaline cleaner of stage 1, for example, from draining into the acid cleaner of stage 3.

Power washers and immersion-tank systems both usually incorporate overflowing weir systems. Figure 4.7 shows a simple tank with a weir. Proper adjustment of a freshwater feed into an ambient rinse causes a slight continuing overflow, which removes almost all of the floating contaminants. The rate of overflow is based upon the percentage of contaminants placed into the water. A good point to start at is 10 percent per hour, or a 100-gallon overflow in a 1000-gallon tank.

Stages 2 and 4 in a five-stage washer system are usually the freshwater ambient rinses; approximately 10 percent by volume are overflowed from each of these tanks every hour. Chemical suppliers suggest frequent dumping of the rinses in order to eliminate possible contamination.

Historically water has been inexpensive in most parts of the country. As usual, we have wasted it. Now we are making an effort to conserve water, even our rinse water. Figure 4.8 shows one way to get your money's worth from each gallon of water by getting multiple use of all rinse water. Figure 4.9 shows fresh water entering stage 4, overflowing from stage 4, and being pumped to stage 2. From stage 2 it can drain to another use; for instance, it can be used to rinse ash from the burn-off ovens that remains on burned-off parts.

To heat or not to heat.   That should not be the question. Millions of dollars have been spent in the development of ambient-temperature

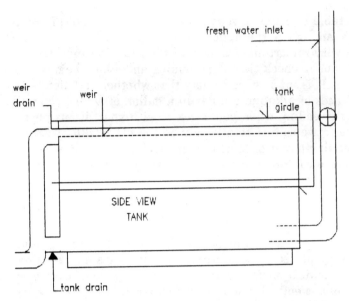

Figure 4.7 Immersion tank with a weir. The water enters in the lower section of the tank; as the water level rises, contaminants float over the weir and out of the tank.

aqueous-cleaning systems for industry, and millions of dollars have been spent in the use of these systems. So tell me why so many people still use heated chemical systems?

I have no quarrel with the chemical industry. They have worked very hard to meet our needs. But the captains of industry are somewhat fickle. Let's start with OPEC. Fifteen years ago, energy costs (gas, oil, and electricity) started to skyrocket. In an effort to allay the complaints of industry, chemical companies started to develop low- and ambient-temperature pretreatment systems. The captains of the manufacturing industry pushed the captains of the chemical industry to the wall, demanding development of the new systems. But as soon as the energy prices started to stabilize, many of them went back to the old systems, leaving chemical companies with large research-and-development costs to pay off. Interestingly enough, most chemical companies have reduced their own operating temperatures as a result of this R&D work.

Water quality

Reverse-osmosis (R/O) processed water. Reverse-osmosis processed water is used when the quality of the water available for pretreatment is not up to an acceptable standard. Your chemical supplier

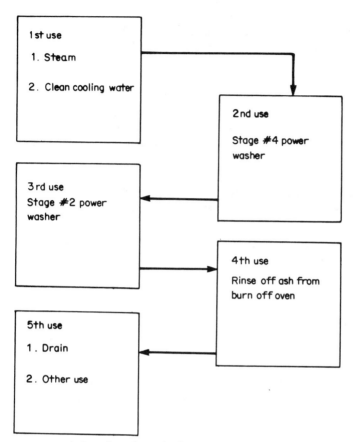

Figure 4.8 A flow chart for multiple use of water in a finishing system.

Drain to other uses

Figure 4.9 A system designed to maximize a water supply. (a) Note the freshwater intakes for stages 2 and 4 and the overflow weir on rinse tank 4 that recycles the water to tank 2; the water is then recycled through another overflow weir to other uses in the system.

will test your water to determine its quality and, if necessary, advise the use of a reverse-osmosis process whereby your water is pumped through a series of filtration banks or chambers to remove contaminants.

Deionized (D/I) rinses. Deionized rinses, or D/I rinses as they are known, are used when and where the water supply itself contains an abundance of contaminants. These contaminants are usually first observed when your chemical supplier runs a water sample test. Solid particles of many names and descriptions will dry and form what I call "fairy rings," white chalky rings or spots, on the surface. These particles are usually thick enough that they will show through a thin one mil cured coating. They will most certainly show on your clean, dried part as little flakes or loose particles. On plated surfaces, they will completely ruin any look you are trying to enhance.

Dry-off ovens

After spending several minutes removing the surface contaminants from your part, replacing the surface with a good phosphate coating, and following with a good seal rinse, what should you do next? You have the following choices:

1. Let the part air-dry, during which time it will immediately start to oxidize and rust.
2. Immediately place the part in a dry-off oven for about 5 minutes at 225 to 300°F.

Of course, the dry-off oven is the most sensible choice; I will not elaborate on the first option. The part should be exposed to the heat immediately upon leaving the washer; good systems will start forced warm air onto the part surface immediately.

Dealing with "puddles" trapped in the parts. Dry-off ovens are usually not designed to remove puddles of excess chemicals from your parts. This liquid should be removed by other methods. I will mention the three most popular:

1. Redesign the hanging fixtures to allow the part to drain.
2. Redesign the part to eliminate the puddle.
3. Remove the puddle at the end of the washer cycle with a suction system, either automatic or manual.

Exposing pretreated parts to excessive heat in an attempt to remove puddles could ruin the pretreatment chemicals on the surface. But the alternative of leaving the puddle on the part will cause your powder to turn to a slurry, which will neither help the appearance of the part nor cure the powder.

Titration, titrating, and titrators

Is a titrator the one who should titrate your aqueous solutions? Yes! Can we then say that, if it's the titrator's job to do the titrating of your chemical solutions and the titrator does not do the job, the titrator is a "ti-traitor" to your company? Most assuredly!

You can spend millions of dollars on your finishing line, but like a chain, the line is only as strong as its weakest link. If that link is the ti-traitor, then in the long run, the ti-traitor will cost you a lot of money. Consider the costs of repairing the improperly pretreated parts that leave your plant.

Your best investment after purchasing your new system is to make sure you get someone who understands the pretreatment system and is well trained to titrate it, add chemicals, run tests, etc. Your chemical supplier will do the training and will help you with the proper titration testing equipment.

Record keeping

It is a very good idea to have a good record-keeping system for your pretreatment operation. Figure 4.10 gives examples of the charts that are available from chemical suppliers. If this type of chart will not do, develop your own. The important thing is that you have one and that it is filled out as often as is necessary. Figure 4.10a shows a typical chemical pretreatment chart for a three-stage iron phosphate washer; Fig. 4.10b shows a typical chemical pretreatment chart for a five-stage iron phosphate washer; and Fig. 4.10c shows a procedure chart for doing the titration procedure.

Shopping for a Pretreatment System

Shopping for a pretreatment process is very important, as you will know by now. You will need to know a lot before you start your shopping. You will need to know the exact condition of the parts you'll be receiving, and about how long they will have been in that particular condition. As an example, let's examine some of the worst hypothetical conditions, since they may be the ones you'll actually encounter:

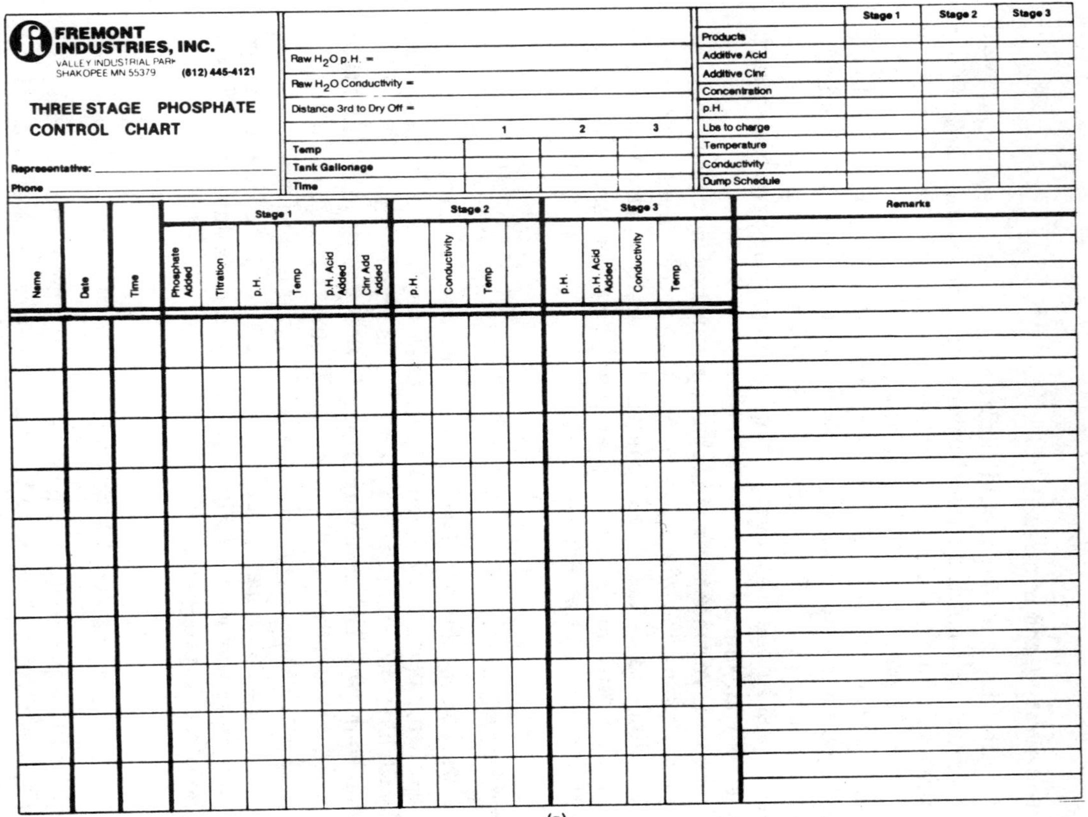

FREMONT INDUSTRIES, INC.
VALLEY INDUSTRIAL PARK
SHAKOPEE MN 55379 (612) 445-4121

THREE STAGE PHOSPHATE CONTROL CHART

Representative: _____
Phone: _____

Raw H_2O p.H. = _____
Raw H_2O Conductivity = _____
Distance 3rd to Dry Off = _____

	1	2	3
Temp			
Tank Gallonage			
Time			

Products	Stage 1	Stage 2	Stage 3
Additive Acid			
Additive Chr			
Concentration			
p.H.			
Lbs to charge			
Temperature			
Conductivity			
Dump Schedule			

Remarks

Name	Date	Time	Stage 1						Stage 2			Stage 3			
			Phosphate Added	Titration	p.H	Temp	p.H Acid Added	Chr Add Added	p.H	Conductivity	Temp	p.H	p.H Acid Added	Conductivity	Temp

(a)

Figure 4.10 (a) Chemical pretreatment chart for a three-stage iron phosphate washer; (b) chemical pretreatment chart for a five-stage iron phosphate washer; and (c) titration procedure chart. (*All samples courtesy of Fremont Industries, Inc.*)

(b)

Figure 4.10 (*Continued*) (*a*) Chemical pretreatment chart for a three-stage iron phosphate washer; (*b*) chemical pretreatment chart for a five-stage iron phosphate washer; and (*c*) titration procedure chart. (*All samples courtesy of Fremont Industries, Inc.*)

	1	2	3	4	5
Low line speed ___ Aug. line speed ___ High line speed ___	Risers ___ Length ___	Risers ___ Length ___	Risers ___ Length ___	Risers ___ Length ___	Risers ___ Length ___

CHEMICALS ___

CONCENTRATION ___

ORIGINAL CHARGE ___

AVERAGE ADDITIONS ___

TIME ___

TEMPERATURE ___

pH ___

P.S.I. ___

CONDUCTIVITY ___

DUMP/CLEAN SCHEDULE ___

COMPANY ___ PHONE# ___ S.I.C. ___

ADDRESS ___ TITLE ___ SALESMAN ___

ATTN. OF ___

(c)

Figure 4.10 (*Continued*) (*a*) Chemical pretreatment chart for a three-stage iron phosphate washer; (*b*) chemical pretreatment chart for a five-stage iron phosphate washer; and (*c*) titration procedure chart. (*All samples courtesy of Fremont Industries, Inc.*)

Situation 1: Say that 95 percent of your product is manufactured in-house; the other 5 percent is made on the outside by a company that uses entirely different steel from an entirely different source.

Situation 2: Say that the subcontractor uses stamping oils, unlike the water-soluble oils your company uses.

Situation 3: Say that the subcontractor uses buffing and polishing compounds on your polished brass parts, and sometimes your parts will sit in his shop for two months before you get them. Will your cleaner still remove these dried, hard compounds, or will you need a different cleaner?

What do you need to know? What do you need to do?

If you really need high-speed production, you'll need a power washer. If your parts are complex, can the spray from the cleaning nozzles reach all the difficult locations? Run some tests. Your chemical supplier will know a shop where you can run parts. Run some and check the results. Is this what you want? If not, try another process.

How much maintenance will your pretreatment system need? Visiting other plants (several of them) and talking to the people who operate and maintain pretreatment systems will give you some first-hand information about maintenance. Be nosy but not obnoxious.

In summary, in shopping for a pretreatment system:

1. You will need to know exactly what kind of process will give the results your company wants and needs *before you buy equipment.*

2. You will need to run tests on a system that is said to give you these results. In these tests, use your product or parts under the exact same conditions you will encounter on your new line.

3. Some of your test parts should be "rushed" to your powder supplier, or preferably to a powder coating operation in your own backyard, within a few minutes after you have completed your pretreatment tests, and powder-coated, if possible with the same powder you will be using. In other words, simulate the exact conditions you will have every day on your own line when you get it into operation. As I have mentioned before, you need more than pretreated panels; you need your actual parts cleaned, phosphatized, sealed, powder-coated, cured, and tested.

4. You will need to know the quality of your water supply.

5

Pretreatment
Equipment

Overview

The pretreatment of your product will vary because of the substrate material, which could be hot-rolled steel, cold-rolled steel, galvanized steel, aluminum, brass, or another metal substrate. The pretreatment will also vary with the methods you employ to manufacture your parts or end product. For example, in many cases where two pieces are welded together, they probably will need some extra pretreatment. The heat from welding will cause a burnt, oily surface, and the burned or carbonized material will be difficult to remove. A mechanical pretreatment system will do a good job in removing this material and smoothing the surface. The system could be an abrasive blasting system that uses a dry or wet material under air pressure. The abrasive material could be sand or another material. It also might be used with or without a liquid, that is, in slurry or dry form. Another method would be to use a centrifugal wheel. In this method the media is transported to a centrifugal wheel which, in rotating, causes the media to be thrown with much force against the surface of the part. Steel shot is frequently used as a media for the centrifugal wheel. A situation where a mechanical pretreatment system is needed is shown in Fig. 5.1.

Another condition you could meet up with in your pretreating process is the simple weld of two thin-gauge pieces of steel. When parts like this are coated and placed in a curing oven, any trapped oils and greases may find a way of leaking out and mixing with the coating, which will cause contamination and probably lead to a reject. A sample of this situation is shown in Fig. 5.1*b*.

Mechanical pretreatment, although not universally used, can be a real lifesaver under certain circumstances; it can be the beginning of a perfect pretreatment system. Proper mechanical pretreatment will

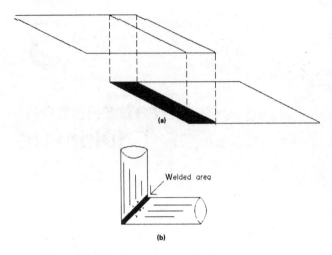

Figure 5.1 Two candidates for mechanical pretreatment. (a) The black section represents an oily, greasy area that is to be welded without cleaning. The trapped soils could later bleed out during the curing process and cause a reject. (b) When hot-rolled steel pipes are welded together, the heat from the process can cause carbonized oils and grease to adhere to the seam; these soils can be removed with mechanical pretreatment prior to cleaning in an aqueous system.

clean welded areas of your steel quickly and efficiently. It will also help give a good smooth surface to your hot-rolled manufactured steel parts. A good chemical pretreatment system following the mechanical pretreatment will give you the best possible surface to apply your powder, providing an acidic surface as an excellent base.

This chapter will deal primarily with the methods and equipment used in the application of aqueous pretreat materials to parts prior to their being powder-coated. As previously mentioned, the application process will vary; people will make all kinds of suggestions as to how you can or should pretreat your parts. I hope that after you finish reading this chapter, there will be no question in your mind as to how to pretreat your parts properly.

Rags, Tanks, Washers, Wands, Ultrasonics, and Ovens

There are several ways you can apply and rinse aqueous cleaning solutions. Let's look at all of them, and discuss them in turn:

1. Cleaning rags

2. Tank cleaning systems

3. Power-washer systems

4. Spray-wand cleaning

5. Ultrasonic cleaning

6. Dry-off ovens

Cleaning rags

Cleaning rags just can't do the job. Through the years, I have watched people valiantly try to remove oil from parts with rags soaked in solvents and various types of aqueous cleaning solutions. The most the rags ever did was to move the oil from one location on the part to another. Even now we occasionally see someone attempt to clean a hidden portion of a complex part, prior to its entering an automatic cleaning system. Maybe this way *some* of the oil the power washer can't reach will be removed, but you'll still have a flawed surface. A spray wand, as discussed below, is the proper tool to use. It will remove oils in remote areas and apply phosphate coatings and seal rinses, giving you the proper protection for your parts, particularly in those remote areas that are too difficult to reach with the power washer's nozzles.

Tank cleaning systems

Systems for steel.   This book is about powder coating. It is also about production, about using powder coating as that final finishing touch in your production. In the previous chapter I alluded to the problems that cleaning tank lines can create if you are interested in production. If you feel you must use an immersion-tank system prior to powder coating, then you can't have a high-speed production line. High speed, multiple-part production systems and tank lines do not go well together because of the labor involved in the handling of the parts.

But if immersion cleaning is best for your parts, then immerse them in a tank system. The immersion of parts can do an excellent job in removing some soils, provided the parts have no pockets in them to trap air and thus prevent the pretreat chemicals from getting to the surface of the part. The important thing is, don't touch the parts and don't let them air-dry between cleaning stages. I have seen systems which provided immersion in aqueous solutions and then used ultrasonics prior to putting the part through a spray washer; these systems worked very well.

Now let's talk seriously about tank lines. The chemicals used in the tanks on a tank line are very similar to those used in power washers. If you're pretreating steel parts, and the soils are not too difficult to remove, you can probably get by with a five-stage line.

Building a set of tanks in itself is not difficult, once you find out what materials you'll need for the chemical system you plan to use. But you will find that buying a tank system is going to be a lot easier and probably less expensive (if you count your labor dollars) if you purchase it from a reliable, experienced source. If you are going to be using a strong acid solution, stainless steel will make your tank last longer; in most cases it's even mandatory. Strong acids will invade welded areas and dissolve them, leaving you with many rusted steel panels lying on the floor of your plant. If you're going to heat the tank and pump the solution, it will take a stainless-steel burner run. All piping and the pump itself will need to be stainless steel.

There are some instances when a plastic tank can be used, but make certain it can handle the heat you will need (if heat is needed). Plastic can also be used for the overflowing rinse tanks. Plastics are an excellent choice if they fit in with your exact needs; they are comparable in price with steel tanks, and they don't rust.

The size of your tank will determine the manner in which it is built. The exact design should be determined by whoever builds your tanks, but there are some features I'd recommend:

1. Good "girdles" or angle-iron braces around the circumference will support the tank when you fill it. Remember, water is heavy, and these tanks hold many gallons of it. Figure 5.2 shows a tank with girdles.

2. Feet under a tank will keep the bottom of the tank off the floor and keep it from rusting out too rapidly. Angle iron works well here. Figure 5.2 shows a tank with angle-iron feet.

3. Your tank must have an overflowing weir. You can run the plumbing from the weir on the outside of the tank, or if the weir is protected from parts and/or part baskets, you can run it on the inside. Figure 5.3 shows two types of tank weirs.

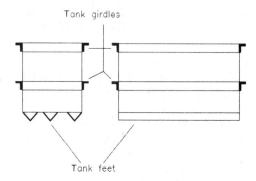

Figure 5.2 A tank with angle-iron girdles and feet.

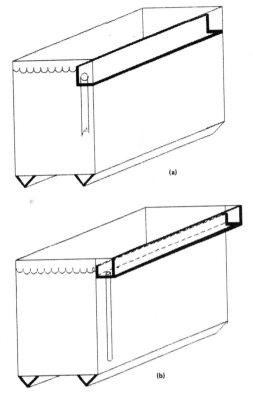

Figure 5.3 (a) A tank with an interior weir and (b) a tank with an exterior weir.

4. The pumping system should be set up to circulate all of your chemical solution. Your chemical supplier may have some excellent suggestions as to how to mix the solution. Figure 5.4 shows how a circulation system and pump can be set up within a tank system. It shows pumping, recirculation, and distribution.

5. The burner run (the tube through which the heated burning gases travel) should be constructed so it will evenly heat your tank. A simulated burner run for a tank is shown in Fig. 5.5. Electrically heated tanks are treated differently. The electricity is fed through rods suspended in the tank. You can see in the figure how heating rods might be suspended in a tank. If you have steam available in your building, it also can be utilized through steam coils. I have seen many tanks with outside wall insulation to help contain the interior heat.

6. A drain is necessary, and the tank should be slightly canted so that when the tank is drained, no liquid will be retained.

7. The drains of all tanks should empty into a common manifold,

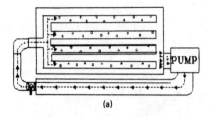

(a)

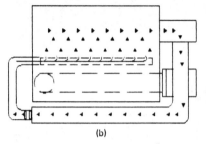

(b)

Figure 5.4 (a) A top view of a circulation system and pump, showing rise layout; (b) pump drawing liquid from the upper levels of the tank through the pump and back into the risers for recirculation. The angles and diameters of the holes in the risers affect the circulation and impingement of the parts. Gas burner and burner run are also illustrated.

and from there, to a holding tank, a drain, or wherever it is you plan to dump your chemicals. See Fig. 5.6.

8. A common manifold for filling tanks is almost mandatory. Figure 5.6 shows an example.

9. A curb, a trough, or a gutter covered with a metal grating surrounding the tank line will help keep liquids within the area designated for cleaning. See Fig. 5.6.

As I have said, tank lines do have drawbacks in plants geared for production. The time and labor involved in using a tank line must be factored into your finishing cost. There are many side effects to consider if you plan to use a tank line:

1. The hands which remove parts from your baskets will contaminate those parts.

2. Floating surface contaminants not removed by your weir will probably be removed by your parts as they are raised from the tank interior.

3. If many small parts are placed in a basket and dropped into a tank, what insurance do you have that those pieces in the center of the loaded basket will ever come clean? And, if liquid does eventually penetrate the area, will it drain properly or will it transfer to the next tank or two, causing contamination?

4. To remove some soils, it does take some form of impingement or collision between the chemical being pumped through the tank

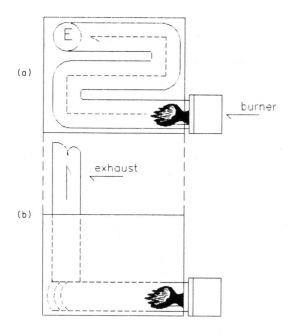

(a)

(b)

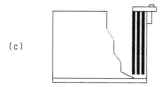

(c)

Figure 5.5 (a) Top view of the burner run in a tank, (b) side view, and (c) side view of a tank with electric immersion heaters.

and your parts. If there is no impingement, there may be no cleaning.

I conclude with some final comments on tank lines for cleaning steel:

1. If your specific parts are large and bulky, with interiors which must be cleaned, use a soak tank or two to remove the contaminants, then hang the parts on a conveyor line, to be put through a power washer.

2. If your parts are large and bulky, consider a spray-wand phosphatizing system. (Read the section below on spray-wand phosphate coatings.)

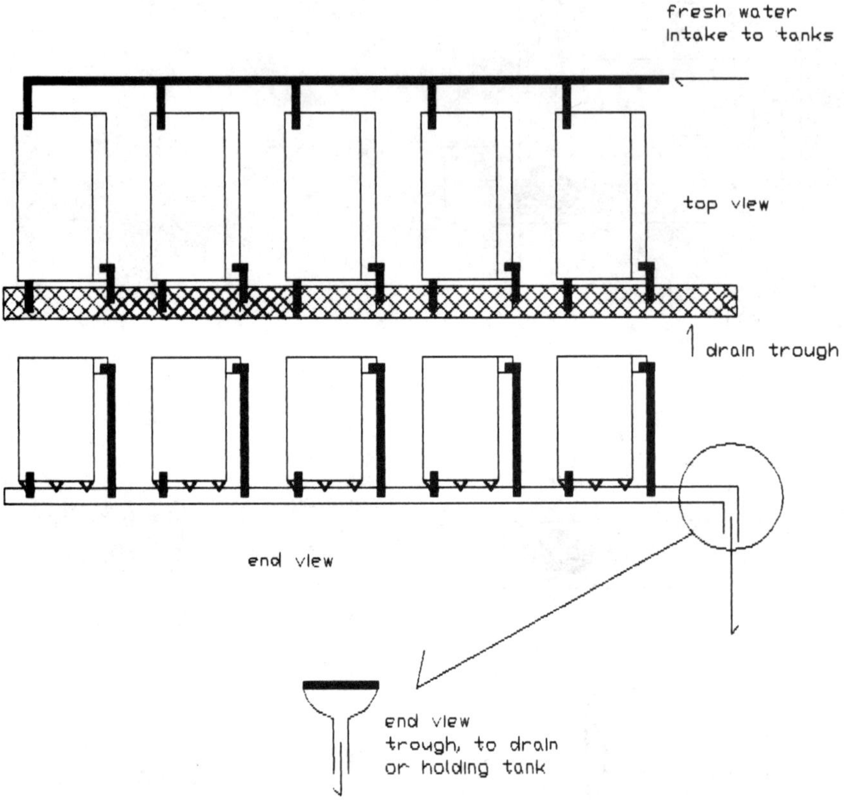

Figure 5.6 Drain systems of a tank line with freshwater intake manifold and common grate-covered trough that goes to a sewer or holding tank.

3. If your only reason for going to a tank line is that you think you can't afford a power washer, put all plans in abeyance until such time as you have enough money to buy a power washer.

Systems for aluminum.   Tank lines work well with aluminum extrusions or fabricated aluminum extrusion parts, particularly if the parts have been welded in the fabrication process. The immersion-tank process penetrates well into the welded areas, and the proper chemicals will do an excellent cleaning job for you. Of course, proper racking must be used so that the extrusions will get the full benefit of the chemicals.

Small aluminum parts present the same problem as do steel parts if they are just dropped into a basket. The chemicals will have problems getting to the parts in the center of the basket. And after they finally get there, they'll have trouble leaving. Aluminum pretreatment usually calls for the use of acidic materials, so stainless steel tanks are a must.

Some extrusion coaters use the same hanging racks for pretreatment as they use for coating. Thus no handling is required, once the extrusions are hung on fixtures. If you're pretreating aluminum, and you want the proper protection, look back through Chap. 4; it will tell you how to pretreat aluminum using a tank line.

Power-washer systems

The power-washer system is my favorite way to pretreat any part prior to powder coating, or for that matter, to pretreat for any organic coating. A properly designed and properly used washer can be your best friend on a finishing line.

How does one design a good power washer? You start out by establishing the conditions for the best pretreatment for your parts. You learn what is the best by repeated testing until you know for sure the best process for your product and product standards. Then you apply these pretreatment steps to the basic washer design.

You can gather a lot of important information before you even talk to a representative of a power-washer company. I list here some things you should know in advance and cite some of the options you'll have when you buy a washer.

1. You'll need to know approximately how many pounds per hour you'll put through the washer, including parts, product, conveyor, and hanging fixtures.

2. You'll need to have some idea of how you'll hang your product so that an envelope, package, profile, or silhouette size can be established. You'll need to know the maximum part width and height, plus the height of the hanging fixture, plus the overall length of the part and hanger in the direction of conveyor travel. Figure 5.7 shows the package and envelope.

These guidelines mean that you'll have to come up with the approximate overall length, width, and height of the complete washer. You'll also have to come up with an approximate pounds-per-hour weight factor. How do you do this? Well, you know the pounds per hour going through the washer. You should know the fixture size and part or parts per fixture. Knowing how many parts per hour will go through the system will tell you the feet per minute of conveyor travel. You know from your chemical testing what chemical process you will use.

Let's establish some parameters and build a washer. Hypothetically, we have established the following system for pretreatment:

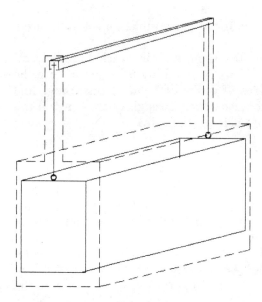

Figure 5.7 Package (solid lines) and envelope (dotted lines).

Stage	Time, min	Process	Temperature
1	2	Alkaline cleaner	140°F
2	1/2	Overflowing rinse	Ambient
3	1	Iron phosphate	140°F
4	1/2	Overflowing rinse	Ambient
5	1/2	Acidic seal rinse	130°F

We know, because of the parts-per-hour rate we need, that our conveyor speed will be 4 feet per minute. We have figured our hypothetical package size as 8 feet in length in the direction of conveyor travel, 2 feet in width, and 2 feet in height for the part plus an additional 2 feet for the hanging fixture space between the conveyor and the part.

Parts are to be in stage 1 for 2 minutes and the conveyor speed is 4 feet per minute:

$$2 \text{ min} \times 4 \text{ fpm} = 8 \text{ ft}$$

Now we determine the figures for subsequent stages. Stage 2 will be 2 feet long. Washer and chemical people don't like stages too short, so let's go to a minimum of 3 feet for stage 2, and likewise stages 4 and 5. Stage 3, a 1-minute stage, will be 4 feet long.

We've established our part package as being 8 feet long. Between stage 1 and stage 2, there must be adequate time for the part package to drain off. If there's not, then the chemicals in stage 1 will be carried over to stage 2, thus quickly reducing the cleaning power of stage 1 and quickly contaminating stage 2. Most washer manufacturers suggest that if you have an 8-foot package (the length in the direction the conveyor travels), then you need at least 2 additional feet to make certain spray-nozzle liquids do not bounce off parts or travel along parts into the next stage. If you cheat at this point, you are only hurting your company. Remember, it costs only a few dollars for sheet metal, while contamination costs, chemical costs, and even disposal costs can be high.

Another item to be considered is the vestibule located at each end of the washer. Its size is predicated on the length of your parts and the line speed. If the leading edge of your 10-foot-long part is being cleaned while its rear portion is still outside the washer, chemicals might travel along the length of the part and make their way to the floor of your plant. The size of the vestibules must also be calculated so as to include the placement of exhaust fans, which provide somewhat of a seal or curtain for the washer and your plant. They keep the moisture-laden heat inside the washer.

Let's start building our hypothetical power washer:

Entrance vestibule, 8 ft long	8 ft
Stage 1, 4 fpm × 2 min	8 ft
Drain for stage 1	10 ft
Stage 2	3 ft
Drain for stage 2	10 ft
Stage 3, 4 fpm × 1 min	4 ft
Drain for stage 3	10 ft
Stage 4	3 ft
Drain for stage 4	10 ft
Stage 5	3 ft
Exit vestibule*	8 ft
	77 ft

*The exit vestibule acts as the drain for stage 5.

Figure 5.8 shows a side view of the 77-foot washer. Figure 5.9 shows an end view of our hypothetical washer with a part in it. You can see the approximate liquid level of the aqueous chemical solution. Note the small opening designated for the vertical section of the hanging

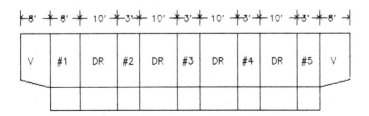

Figure 5.8 Side view of hypothetical power washer.

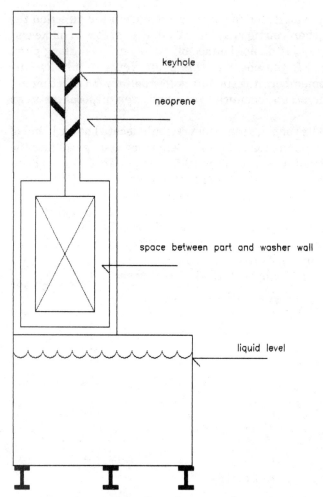

keyhole

neoprene

space between part and washer wall

liquid level

Figure 5.9   End view of hypothetical power washer, with a part in it.

fixture, between the conveyor and part. This is sometimes called a *keyhole,* and it is possible for some stray chemical and/or moist warm air to enter this area. A number of methods have been introduced to help prevent that, including the use of neoprene seals. Pressurizing the area can also help. The important thing is to keep as much chemical and moist heat off the conveyor as possible.

You will also see in Fig. 5.9 that there is some space between the product and the profile of the washer. This opening is itself called a *profile* or a *silhouette.* Different manufacturers leave a different space. To some extent the weight of your product will have a bearing on the size of the space; usually it will be somewhere between 6 and 12 inches on each side, with the same space on the top and the bottom.

In the upper section of the washer of Fig. 5.9, you'll see the profile or silhouette of the part going through the washer. This same profile or silhouette should appear throughout the washer, at the beginning and end of each stage. The silhouette appears as a separation between each stage. The reason for this particular type of silhouette is to help keep spray-nozzle chemical deflection down to a minimum in order to keep the carryover of spray chemicals to a minimum.

The drain section between stages is meant to collect the chemical solution that drips from the part and send it back to the area or tank where it belongs. Going in the direction of travel, the first two-thirds of the drain section is angled to drain toward the stage just completed; the last third is angled to drain materials toward the stage that will be entered next. Figure 5.10 shows how the drain system helps to keep the materials in their respective tanks as the part progresses through the washer.

No matter, there will be some chemical splashing. One way or another, liquid will get on the floor of your plant. Thus it's a good idea to elevate the washer tanks slightly from the floor on which they sit. It's also a good idea to surround the area with a small trough and over that trough to place a grating.

Have the trough empty into a drain. This will keep most of the chemical spill from the rest of your plant.

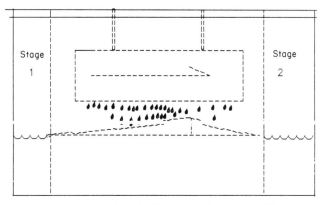

Figure 5.10 Drain system. As the part exits stage 1 of the washer, excess aqueous chemical solution drains off, falling onto an angular drain board. The first two-thirds of the drain board drains back to stage 1; the remaining third drains toward stage 2. The nozzles will start to spray the part as it enters stage 2; it is this spray that will drain back into the stage 2 tank. To help prevent excess nozzle spray into the drain area there are baffles that deflect the spray back into their respective stages.

How and where will the washer be placed? Can it be a straight-through type washer? Or, because of your building size and layout, must there be a turn? Or, will it be necessary to have a U-shaped washer? That's a decision for you to make.

Another important point: How long will it take a part to get from the washer to the dry-off oven? Remember, oxidation will start if parts are left out in the open air with water, or even a mild acidic solution, on them. Under normal circumstances, there will be an elevation drop in your conveyor from where it is located inside the washer to where it is located on the ceiling of the dry-off oven. Depending upon the length of your product, this distance could be considerable. The actual drop should be kept to a minimum of about 4 feet. If your parts are 8 feet long, a shallow conveyor drop will be necessary, giving oxidation a good chance to get started. Many companies eliminate this potential problem by "pitting" the washer, that is, placing it in a pit. Then, when the parts leave the washer, they immediately enter the dry-off oven, since it's not necessary to change conveyor elevation. The pit must be large enough to service the washer as necessary, must have drains, and must also be able to contain overflowing water, splashes, or solutions when necessary. Figure 5.11a shows a washer at floor level, and Fig. 5.11b shows a washer in a pit.

It's a good idea to equip your washer drains with a common manifold so that all drains exit at one location. Then you'll have complete control over the destiny of any solutions leaving your washer. I have seen situations where it was discovered, when retrofitting a plant with a finishing system, that there was no drain anywhere within the

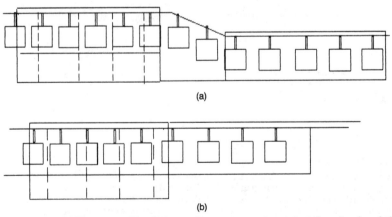

(a)

(b)

Figure 5.11 (a) Parts traveling between the power washer set at floor level and the dry-off oven; (b) the same scene, but with the power washer set in a pit. You can see the problem with oxidation introduced by the air space between the two units in (a) and the elevation differential between the washer and the oven.

vicinity. Figure 5.12 shows how we handled the problem. All liquids leaving the washer were drained to the sump through a common manifold. From the sump, the liquids were pumped through plastic pipe to the ceiling of the plant. From there they went either to a treatment system or to the regular drain system of the building.

Let's continue with the power-washer design. Each washer stage will supply its own chemical solution, but it will also have to accommodate some mechanical action. For one thing, each stage will contain a heat source if the solution requires heating above ambient temperature. As illustrated in Fig. 5.13 it will also contain:

1. A clean-water entry (A) split into two lines, one line for automatic trickle filling as contaminated solutions and floating soils leave through the weir and a second line valved for quick filling of the tank when it has been emptied

2. A pressure gauge (B) to monitor the pressure being sent to spray nozzles by the pump

3. A throttling valve (C) to reduce pump pressure to spray nozzles in order to prevent the removal of light parts from hanging fixtures by the high pressure of the spray

4. Pumps (D) to propel solutions through risers and nozzles

5. An overflowing weir or dam (E) to enable floating contaminants to be removed from tanks

6. A drain (F) for the weir to allow separation of contaminants such as floating oils and greases from the chemical solutions in the tanks

7. Grating (G) over the floor drains and/or gutters that receive all overflow solutions from the various stages and send them to the sewer system or holding tank

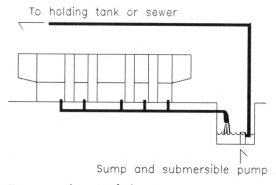

To holding tank or sewer

Sump and submersible pump

Figure 5.12 A remote drain system.

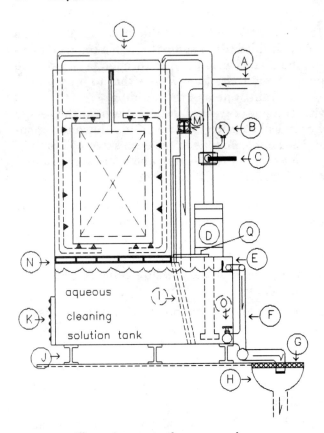

Figure 5.13 The various parts of a power washer.

8. A drain gutter or trough (H) that runs the length of the washer

9. A screen (I) to keep large parts, nuts, and bolts from entering the pumps

10. I beams or channels (J) to keep the washer base off the floor, thus helping to keep spilled or overflowed liquids from starting a premature rust problem on tank bottoms

11. A tank opening (K) so the interior of the tank can be cleaned and heavy concentrations of contamination, solid soils, and small pieces or parts removed

12. Sections of piping (L) to transport chemical solutions to the nozzles that spray the parts

13. Shutoff valves (M)

14. A grating (N) on the top of the tank to prevent parts from falling into the tank and also to permit people to walk through the washer without falling into the solution tank

15. A tank drain (O) for draining the solution when needed

16. A cover (Q) on the tank so that chemical solutions can be placed in the tank (Some washers are equipped with automatic chemical replenishing systems; these units are not shown in the drawing.)

The area in the base of the tank that holds the solution will also contain the heat source. Figure 5.14 shows what is called a *gas burner tube,* or a *burner run,* one of the many designs used to supply this heat to the liquid.

An essential element in the design of the washer, if your manufactured product is covered with machining oil, or any oil, is an *oil skim-*

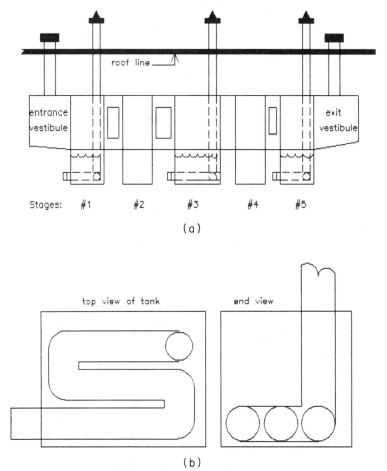

Stages: #1 #2 #3 #4 #5

(a)

top view of tank   end view

(b)

Figure 5.14 (*a*) Three burner runs in a power washer, with vestibule fans and exhausts going from the washer through the roof. (*b*) Enlarged views of the burner runs.

mer, usually placed in stage 1, to remove the oil and to prevent it from contaminating parts you are trying to clean. The first stage of most washers also contains an alkaline cleaner, which is designed to remove organic soils, including oils. As you know, oil and water do not mix; the oil tends to float on the surface of this first-stage tank. Much of it can be removed with a built-in oil skimmer. There are commercial skimmers on the market, and some washer manufacturers make their own to fit their own washers. The skimming of the oil will help add to the longevity of the aqueous chemical solution in this first stage.

Maintenance access doors are important for the internal check of any washer. They are usually installed at the drain-off section of the washer. A weekly inspection of spray nozzles is easy when you are able to view the operating stages of your washer through an access door.

Washer lighting is an essential item. If you have ever been around a power washer, you know there are many occasions when someone will need to enter the washer. Hanging fixtures sometimes break, people hang parts the wrong way, and normal maintenance requires the occasional changing of nozzles. Whatever the situation, it's difficult if not impossible to work inside a washer, using only one hand to do the work while holding a flashlight with the other. But if dock lights, such as are usually mounted at the loading dock of a building to illuminate the interior of a truck being loaded or unloaded, are hung outside the washer next to each access door, they can be swung around into the washer through the access door opening, allowing you to view the interior operation of your washer.

"Clean outs" are the covers, usually rectangular-shaped, that are mounted on the side of power-washer tanks. They are usually bolted very tightly to the lower side wall of the solution tanks. After the tanks are drained of chemical solutions, an inspection will usually show substantial quantities of solid sludge remaining. This sludge can be hosed out or shoveled out by removing the clean-out door. The doors are frequently sealed with a neoprene gasket. Surprisingly enough, I once saw a ship's hatch cover utilized as a type of "quick-removal" door. A couple of wing nuts were used to seal it.

Typical maintenance doors, dock lights, and a clean out are shown in Fig. 5.15.

I have seen many washers over the past years. The ones which have been the most leakproof were those which were manufactured in the plant of a company that was in the washer business. You could tell that the company had taken plenty of time to weld the upper walls of the washer where the parts are spray-cleaned and where a perfect seal is essential. If it is at all possible, have the manufacturers build this

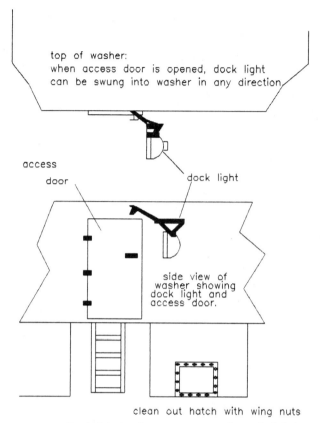

top of washer:
when access door is opened, dock light
can be swung into washer in any direction

access
door

dock light

side view of
washer showing
dock light and
access door.

clean out hatch with wing nuts

Figure 5.15 Typical design from maintenance doors, dock lights,
and clean out.

upper section in their own plant rather than in yours. When this isn't
possible because of the overall height of the washer and the upper section must be built in your plant, don't let your own people do the welding job. Have recommended professionals do it.

Lights, bells, whistles, pumps, nozzles, gauges, and other gadgets. Lights,
bells, whistles, pumps, and gauges make up a major portion of your
washer, a very important portion.

Let's get the pumps and nozzles out of the way first. I have known
pumps to operate for many years; I have also known pumps to fail
soon after they were installed. I have been in many factories which
were located in towns where the closest pump replacement was 1 or 2
days away. An old friend once told me that he protected himself from
being without a pump by always having a spare one in storage in his
plant. He also specified that his tank be designed so that all washer

pumps purchased and installed be of the same size—for example, that small stages use a 5-horsepower pump and large stages use two 5-horsepower pumps. (The horsepower rating of the pumps in your washer will be based on many calculations, which I won't detail here.)

From the pump, your chemical solutions go through an assortment of pipes called "headers" and "risers." From the risers the solutions go through nozzles especially designed to spray your parts. Nozzles are made from steel, stainless steel, and an assortment of plastics. There are metal nozzles which must be removed with healthy pipe wrenches when they are plugged or need replacement, and there are quick-disconnect nozzles which snap on and off. Nozzles are made with all types of spray patterns. Among available patterns there are flat spray nozzles and hollow cone nozzles. Nozzles spray at different angles, varying from a minimum of 30° to somewhere in the high-90° range.

One of the newer popular devices for the power washer is the clip-on type plastic spray-nozzle assembly. If you have ever tried to remove rusted steel nozzles from a power washer, you will know what an improvement these are from the old screw-in type nozzles of the past. The nozzle housing itself clips on to the washer riser, as shown in Fig. 5.16a. The nozzles themselves are available in styles to do any normal job. Should you still wish stainless steel nozzles or any other metal, they can be screwed into the ball assembly that sits in the cap of the unit. Figure 5.16b shows the interior of a power-washer section with these units.

Among the gadgets available for power washers are those that regulate the amount of pressure the spray nozzles direct at the parts. The impingement of solutions is very important. You cannot expect small, lightweight parts, 100 of them hanging from flimsy hooks on a light-weight hanging fixture, to be sprayed with the same pressures as your 100-pound parts. Usually mounted on the side of your washer is a pressure gauge and a throttling valve to control these pressures; be sure someone is responsible for adjusting them when necessary.

There are also a number of systems available for automatically feeding chemical solutions to the various stages of your washer. The power-washer control cabinet comes equipped with many lights, bells, and switches dedicated to the operation of each stage. It is even possible to start your washer in the morning with the aid of a programmable controller that controls the entire process. You can imagine the devastation that would occur if someone interrupted the cycle and shut off your conveyor line with parts in the washer. Many years ago, I bought a new American-made car whose paint began to peel within 30 days of purchase. It was obviously a case of a faulty finish process. I still buy American cars, but of another brand, and I don't know whether the other company has solved its problem yet.

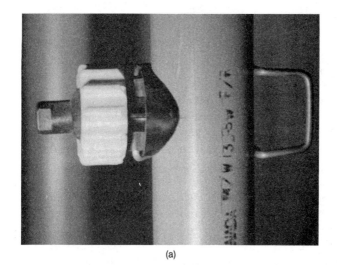

(a)

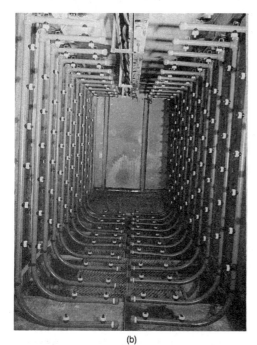

(b)

Figure 5.16 (*a*) Clip-on plastic spray-nozzle assembly and (*b*) interior view of the washer with the clip-on unit. (*Both photographs courtesy of Bex, Inc.*)

Spray-wand cleaning

Because of the design, there are some products with parts that just can't be reached by the nozzles of a power washer. If you say to yourself, "Forget it," you're making a major mistake, especially since there's an easy answer for your problem: a spray-wand cleaning and phosphatizing machine. Those places hidden from the spray-washer nozzles can probably be reached by a spray-wand phosphate machine that cleans, phosphatizes, and seals those hard-to-get-to places. The machine must be able to handle the temperatures you need for the chemicals, be able to inject the proper percentage of chemical, and also be able to handle the pH ranges you will require for the alkaline cleaner (12 + pH), down through the phosphate and final rinse (close to 3.5 pH). Elsewhere in this book (see Fig. 4.2) is the photograph of a stainless steel spray-wand machine that has three injector systems for handling stages 1, 3, and 5 of an iron phosphate cleaning system, along with the rinses. A machine like this can clean the hidden areas of a part prior to putting it through a proper power-washer cleaning system.

Ultrasonic cleaning

A few years ago, I met with a difficult cleaning problem. We were trying to clean many small parts pitted with a dried organic-based material. This material was known to be extremely difficult to remove. But with ultrasonic testing using a heated solution, we were able to remove it completely within seconds. Ultrasonic cleaning is the fastest- acting agent for cleaning purposes that I have seen in a long time. Remember that ultrasonics used properly with aqueous cleaning solutions will give amazing results. But also remember that you'll have to have a method of removing large quantities of contaminants from the ultrasonic solution quickly or you will be changing the solution frequently.

Dry-off ovens

Yes, the dry-off oven is actually a part of the pretreatment system. Why? Tell me what happens when you put water on a clean piece of steel or aluminum. The water will almost immediately start the oxidation process of the steel or aluminum. The oxidation will give you a surface no coating will adhere to properly. If you simply apply powder over the oxidized portion, the oxidation process will continue; eventually the surface coating will fracture and further expose the substrate to the atmosphere.

Once you have started the finishing process, with the cleaning, phosphatizing, and sealing of your parts, dry them as quickly as you can with heat. Then powder-coat and cure them. Remember Lehr's Second Law: Don't clean today what you cannot immediately powder-coat and cure.

6

Powder Application Equipment

Overview

The application of powder is not a difficult process to understand once you know how each piece of application equipment works. You probably have some vague idea about the theory of electrostatic powder application with a spray gun. But if your management people ask you how to reclaim powder, what can you tell them? What in the world is a fluid bed? Believe me, it's nothing like a water bed. Then there is always the "booth," or the spraying enclosure, and all the other components of a powder application system. I hope that when you are finished with this chapter, you'll be able to answer any questions management may ask about electrostatic powder spraying, tribocharging, powder handling from the carton, reclaiming of powder—and all the other questions that arise when a new powder coating facility is discussed.

The spraying of powder electrostatically is nothing like the spraying of liquid coatings, either with or without the help of electrostatics. Spraying powder is not like spraying the fender of an automobile with nonelectrostatic wet paint. The wet-coating spray strokes do not resemble the "floating" of charged dry-powder particles as they journey from the gun to the article being coated. The floating particles with their electrostatic charge will cover a broader expanse of surface than a nonelectrostatic wet coating will. I frequently have seen people, who before had only sprayed nonelectrostatic wet coatings, coat with powder. They invariably make many more strokes than is necessary to coat a given part.

The beauty of using powder application equipment becomes evident just as soon as it is set up to coat a particular job with a particular powder. Once set up, the equipment will usually operate consistently, and with very little maintenance other than the occasional blowing off of the gun tips which tend to collect some powder during the spraying cycle.

There is a wealth of application equipment available on the market. Obviously, there are differences in design from brand to brand. Equipment manufacturers each want to control a portion of the market, so they are constantly trying to improve their products. You should have seen some of the application equipment offered 15 to 20 years ago; then you could really appreciate the advances in design and safety available to you today.

Your specific wants and needs are going to be different from those of people in other industries, so naturally there will be some differences in the specific equipment you will be considering. Don't be shy. Ask whether or not any of your competition is using the brand of equipment you're looking at. How do your needs differ from those of another company? Do you use one color, two colors, five, or twenty colors? How often will you change colors? Do you have an unusual number of deep recessed areas in your parts which will create Faraday cages? There might be many reasons why one type of system will work better for your company than another. Make certain your prospective supplier is aware of all your requirements. If you don't tell your supplier now, it may be too late after the system is installed.

Let's look at the common components used in the application of powder, at how they function, why they function, and how they work together in the total application picture.

The component groups are the powder hoppers, pumps, and hoses; the charging systems; the spray guns, automatic and manual, their nozzles and tips; control panels; the actual application enclosure; and the collection or reclaim system.

Powder Hoppers

The powder hopper is a container from which you control the destiny of your powder. It is for the short-term storing of powder, prior to its use in the system. It can be round or rectangular, depending upon its manufacturer. If it must be moved frequently, a hopper should be equipped with casters. On the other hand, if it's to be in a fixed location in the system, it need not have casters. The powder hopper will have a system for fluidizing the powder in its base. A bulk dry material like powder must first be fluidized or suspended within a stream of air in order to be transported from one location to another. The system in the base of the powder hopper that causes the fluidizing is called a *fluidizing bed*; it consists of a porous membrane, usually a sheet of plastic material, through which air is passed at a controlled rate. As this air moves through the porous membrane, it mixes with the powder. This fluidizing (the mixing of the air and powder) expands the powder stored above the membrane. Once it is turned on, I like to see a fluid bed operating all of the time. Keeping the powder constantly fluidized with clean, dry, compressed air keeps it from becoming agglomerated, or lumpy. The fluidized powder is thus ready to be used whenever it is needed within the system. Figure 6.1 shows how a fluid bed works.

To see the effect of fluidizing, run your hand through some powder at rest in a fluid bed. It will feel heavy and dense. Now apply some pressure-controlled, dry, compressed air. Give the air a chance to mix

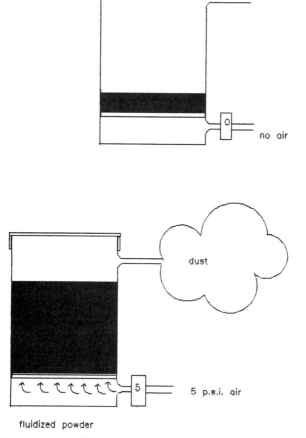

fluidized powder

Figure 6.1 A fluid bed and how it works.

with the powder. The first thing you'll notice is that the powder level will rise to almost double what it was without the air. The powder will look and feel almost like a liquid. Run your hand through it. Isn't it much like running your hand through water? Part of the now-fluidized powder will want to escape from the hopper because of the added air pressure, but the cover on the hopper will keep the powder and dust inside while the excess compressed air escapes through a hose in the upper section of the hopper, along with some powder dust. The air and dust must be expelled into a safe place; the powder coating enclosure makes an excellent unloading location for this excess compressed air and dust.

There are many places where a fluid bed and hopper can be utilized in a system. The more work a hopper does, the less one needs to be concerned with powder contamination. To illustrate, I'll give some examples of powder hoppers and their uses.

1. A remote hopper can hold virgin powder until it is needed somewhere in the system. Such a hopper can be used to supply the gun-feed hopper or to feed virgin powder as an additive to a batch of reclaimed powder. The powder from the gun-feed hopper goes to the coating enclosure via spray guns. Overspray goes to the reclaim system hopper where it is transferred to the overspray holding hopper. Figure 6.2 shows how such hoppers could work together.

2. A feed hopper can be actually attached to the bottom of a spraying enclosure to catch some of the overspray and then feed powder-gun pumps. Figure 6.3 shows this simulated system.

3. A hopper can be attached to a reclaim system and used to help remove oversprayed reclaimed powder. It does this by fluidizing the powder so that it can be pumped back to a feed hopper to be mixed with virgin powder and resprayed onto the parts. Figure 6.4 shows how this would work.

4. A storage hopper can hold reclaimed powder that will eventually be mixed with virgin powder and then sent on to a feed hopper.

Powder Pumps

Powder pumps are used to transport powder to the various locations where it might be needed within the system. Powder is easier to move by pump than by any manual method, and it is usually the more efficient way to do it. We have talked about fluidized beds, and you have seen how we fluidize powder. Unfluidized powder can be pumped, but most systems are set up to pump fluidized powder. Figure 6.5 shows the pump, hopper, and gun working together.

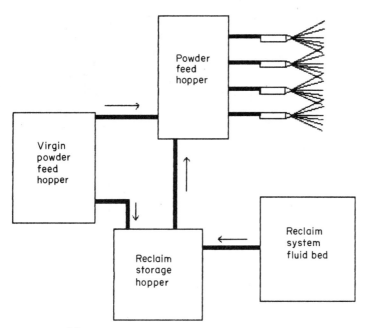

Figure 6.2 A hopper system.

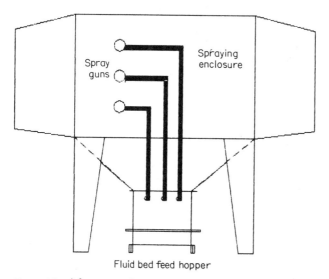

Figure 6.3 A hopper system.

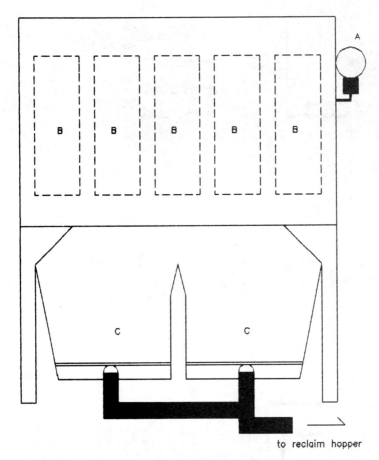

Figure 6.4 A hopper system, including the cartridge purging system (A), cartridges (B), and chambers (C) where the purged powder is fluidized and pumped to the reclaim hopper.

Most of the gun manufacturers have developed powder pumping systems based on the Venturi principle; these pumps vary in design from manufacturer to manufacturer, and I have seen many varieties of powder pumps. The interior of a simulated pump at work is shown in Fig. 6.6.

The Hoses

Different types of hoses are used for different purposes. As a dry, bulk material, powder is slightly abrasive. When it travels through the application system, it will cause some wear on the hoses as well as on other portions of the application equipment. Manufacturers supply

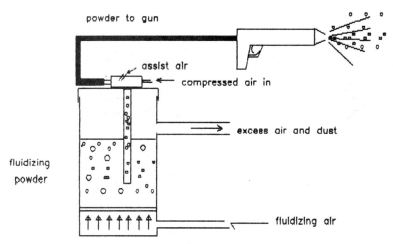

Figure 6.5 Powder hopper, pump, and gun.

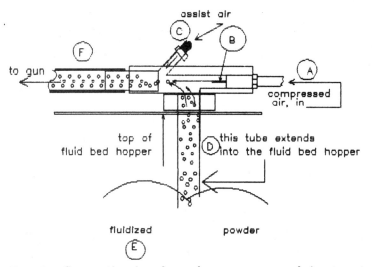

Figure 6.6 Cross-section view of a powder pump: compressed air enters at A; the velocity of the air is increased as it is forced through the restriction at B; as the air passes through the pump to the exit at F, it passes over the tube at D, which contains fluidized powder. This creates a suction, forcing the powder from the tube up into the airstream. This powder is immediately replaced in the tube by more powder from the fluidized bed, E; as the powder in the airstream heads for the exit of the pump, it is further assisted by air entering at C.

hoses which will last for a long time if they are used properly. Hoses are made to carry powder; they are not made to be left on the floor where they can be stepped on or tripped over, nor are they made to be run over by fluid-bed casters. In short, they are not made to sustain the kind of abuse that will cause fracturing on their interior walls. There is an old saying about a chain being as strong as its weakest link. A strong hose can be made very weak by a fracture on its inside wall.

Charging Systems

There are two methods by which the powder is charged in the powder spraying process: electrostatic charging and tribocharging.

Electrostatic charging

Electrostatics is the most common method of applying powder to a substrate. An electrostatic system consists of an electrical charging apparatus and a spray gun through which both the powder and the charge are expelled. The charging transformer is frequently, but not always, housed close to the spraying enclosure. One manufacturer has its transformer built into the gun barrel. The transformer consists of an input side, usually of 110 VAC 60 cycles; the output side yields a high voltage and a very low amperage, measured in terms of microamperes. The electrostatic charge is transmitted by one of several means to the tip of an electrostatic powder-spraying device called a spray gun. Using a hand gun, the operator "triggers" the gun by depressing an electrical switch. In the case of automatic guns, the triggering is done by turning a switch on, usually at a console.

Figure 6.7 shows the sequence of events that occurs when the switch or gun trigger is actuated: (1) Powder is pumped from a fluidized bed, (2) through the venturi pump, (3) through the hose, and (4) through the electrostatic gun. The powder, while traveling within the hose, consists of particles carrying both plus (+) and negative (−) signed charges. (5) As the powder is expelled from the gun, (6) it passes through an invisible ionizing field, (7) which is transmitted from a metal needle at the exit of the gun. (8) The negatively (−) charged powder now travels toward the clean, well-grounded part that's to be coated. The part, being grounded, acts like a receiving magnet for the charged particles, which cling to the clean surface much like a steel nail to a magnet. As the particles cover the surface, the "magnet" effect gradually becomes weaker and the particles look for new clean areas to be attracted to. The force is so strong that the particles will wrap themselves around corners and cling to locations the gun cannot even "see." The powder is assisted in its journey from the gun tip to

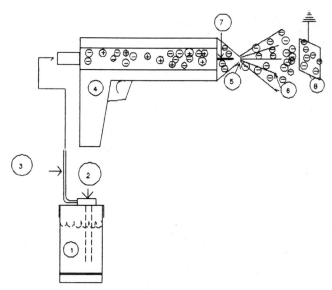

Figure 6.7 A powder gun in operation: (1) fluidized bed and hopper, (2) venturi powder pump and withdrawing tube, (3) feed hose, (4) spray gun, (5) powder leaving gun, (6) invisible ionizing field, (7) ionizing tip, and (8) grounded part.

the part with some extra air, usually introduced to the powder somewhere in the gun. This air is sometimes called "assist air," and also "forward air." Whatever it is called, it helps the powder get to the part.

It is the electrostatics effect that causes the powder to be attracted to the part. The electrostatic relationship between the powder particles and the substrate is based upon the voltage setting. Too low a setting will reduce or eliminate the electrostatic effect; too high a setting can cause other problems, including an increase in the Faraday cage effect (to be explained below). Your specific voltage setting will be dependent upon many variables, including the part configuration, the type of powder, the spraying equipment, and the available ground. The electrostatic phenomenon will do much of your work, if you let it do so. I say this only because I have seen the work of spray painters who are used to spraying nonelectrostatic wet coatings.

Tribocharging

Tribocharging is another method used to charge powder. It is accomplished by mechanical means rather than by a charge received through an electrical transformer. It is accomplished by propelling powder particles through a powder hose and gun. Somewhere within the gun,

Figure 6.8 Tribocharge gun. The powder, propelled by air pressure, enters the rear of the gun; the particle sizes vary and so too does their electrical charge. As the particles enter the gun, they encounter an obstacle that causes them to rub against the interior wall of the gun; this rubbing creates the friction that gives the particles their positive charge.

through one of several methods, the flight of the powder particles is slightly diverted. This diversion causes the particles to touch the interior wall of the gun. The rubbing action causes a friction, or tribocharge. Figure 6.8 shows a hypothetical tribocharge-type gun in the process of charging powder. The drawing shows the plus- and minus-charged powder particles entering the gun, receiving their friction charge (positive) and then being expelled from the gun.

The rubbing action within the gun is similar to the action of rubbing a balloon rapidly on a piece of cloth. The balloon takes on an electrostatic charge and will stick to most surfaces. A similar action takes place when one walks on a carpet during a low-humidity winter day. The body builds up a charge, which is discharged very quickly when you touch something or someone who is grounded.

There is no electrical transformer in use in tribocharging. The charge is accomplished totally within the gun. The result is that most individual powder particles receive a positive charge from within the gun, not from an ionizing field outside of the gun. Thus in most cases, the Faraday cage is eliminated.

Relative merits of electrostatic charging and tribocharging

There are, of course, both advantages and disadvantages to electrostatic and tribocharge guns, and opinions vary from gun manufacturer to gun manufacturer, though many manufacturers offer both types of gun for sale.

Proponents of tribocharging cite (1) the simplicity of operation, (2) the evenness of the coating, (3) the need for less equipment, and (4) fewer problems with the Faraday cage.

But opponents of tribocharging also make their points:

1. Because of the tumbling action of the powder around the diversion area within the gun, there is much wear to the interior of the gun.

2. The charging rate is lower and slower; thus to achieve higher line speeds, more equipment is needed to spray parts than is required of conventional guns. (Powder manufacturers are quick to mention that there are a number of additives which can be added to powder to assist with the charging rate.)

3. Equipment use is affected in areas of high humidity.

4. There are problems with spraying polyesters.

My personal opinion on the subject has always been that if there is an opportunity to give the end user a gun that will do a more efficient job, industry needs it and deserves it. On the other hand, the gun and process must be demonstrably more efficient and economical than equipment presently used if the conversion is to be justified.

Spray Guns: Automatic and Manual

The spray gun is where it all happens. If your pumping system gets the powder to the gun, and the powder gets its charge, the rest is easy. The charging mechanism you use, with the proper adjustment of gun controls, will spray powder onto your parts with very little effort on your part.

Powder sprays so easily and thoroughly, you will want to take advantage of automatic spraying equipment since it ensures the most efficient production levels. But there will always be times when manual reinforcement or touch-up will be needed. Unfortunately, engineers have not been able to design all products to be coated automatically. Manual guns definitely have a place in powder coating. They can reach places automatic guns cannot "see." Manual guns work best in short, simple runs of production, where frequent color changes are the norm, and when the line is going slow enough for the operator to do the job. Manual guns also work well as laboratory guns and for doing sample parts.

Automatic gun systems, on the other hand, consistently produce large quantities of good parts, day in and day out, when they are set up and maintained properly. One operator can operate many automatic guns, whereas it takes one person to operate one hand gun. Automatic guns also provide the consistency of well-coated parts, if they are set up to do the job properly. Consistency is not what you get with a manual spray gun; people have "high" and "low" periods during the day, and the mil thicknesses of manually sprayed parts will vary during these periods.

Automatic guns cannot "see" what they are coating; they will coat the one part in front of them, or one thousand parts, it makes no difference. And they will operate consistently for hours without much maintenance except that required every couple of hours to blow off powder collected in the vicinity of the gun tips.

The important thing in spraying is part presentation, whether the spraying is manual, automatic, or a little bit of both. The part must be presented properly both to the "blind" automatic gun and to the operator of the manual gun. Automatic guns can be controlled in many different ways. They can be placed in a fixed position where they do not move except when manually adjusted. They can be attached to a cross-arm reciprocator, which will propel the guns up and down within an adjustable given range. Several guns can be attached to swivel devices and moved up and down with a single rod in an oscillating mode. Each of the devices have both good and bad points, depending on what you are trying to accomplish. The specific job you want to do will dictate the device which will work best for the job. Figures 6.9 and 6.10 show some of the popular forms of gun movers.

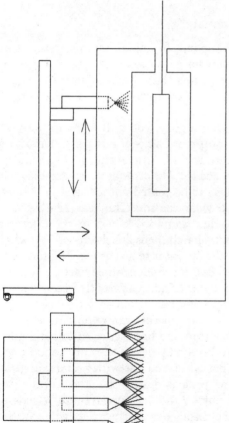

Figure 6.9 Top and side views of a gun reciprocator: the reciprocator is floor-mounted, capable of being moved closer to or farther away from the spray enclosure as required; each gun can be adjusted for height, velocity, and distance by the operator.

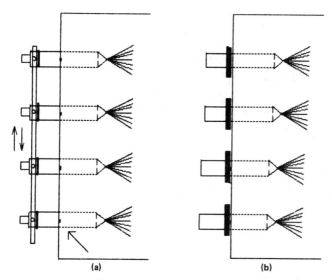

(a)   (b)

Figure 6.10 (a) Oscillating automatic spray guns, which can be moved in or out, up or down at controlled speeds; (b) "fixed" guns which must be individually and manually adjusted, in or out, up or down.

Nozzles and Tips

Automatic guns and manual guns have a variety of nozzles and tips to control the spray pattern they emit. The idea is to change the spray pattern emitted by the gun. Certain tips are used for "flooding" large, flat areas, others for doing the interior of parts. Practice and education will help you determine which will work best with your parts. For manual guns there are tips which extend the end of the gun some 12 to 30 inches, which helps when long reaches into large parts are necessary.

Control Panels

Your powder spraying system will be controlled at a control console. Most companies place this console at a location close enough to the spraying enclosure so the operator can make adjustments to any gun and immediately see the gun's reaction or pattern. Profit-and-loss differentials are set at this location. It is important that adjustments be made properly and promptly, and equipment companies have recently introduced equipment which will help you in this direction. Each gun usually has its own control panel on a single console. Controls vary from company to company, but generally they affect powder delivery volume, air pressure to assist the powder delivery, air pressure to control the spray pattern, and voltage. The console is also equipped with on/off switches, of course, and sometimes a light. Adjustable functions are usually accompanied with an appropriate gauge. Various control panels are shown in Fig. 6.11.

(a)

(b)

Figure 6.11 Various types of control panels. (*Courtesy of Volstatic, Inc.; Thorid Electrostatic Powder Systems; Nordson Corporation; and Ransburg-Gema*)

(c)

(d)

Figure 6-11   *(Continued)*

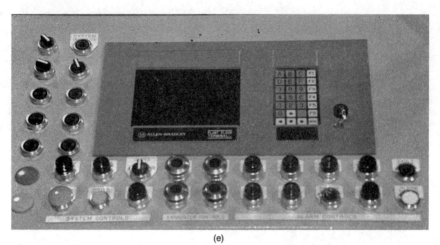

(e)

Figure 6-11 (*Continued*)

Special Considerations

There are some possible problem areas to be concerned with when choosing your powder application equipment. It might do us well to pause here and, based on what we have discussed so far, look at these factors. The problem areas I'm talking about are impact fusion, the Faraday cage effect, and back ionization.

Impact fusion

Powder, being a bulk material and not a liquid, does not navigate 90° corners too well, so it's necessary to use a rounded sweeping curve in all hoses that are laid around corners. Unnecessary restrictions may tend to impede powder flow and cause impact fusion within the hose. *Impact fusion* occurs when the powder being moved under air pressure through the system becomes restricted in one area. This area might be a sharp corner or any other restricting location such as sometimes exists at the exit of your venturi, within the powder pump itself, or in the powder gun. The force of the compressed air against the powder particle will actually fuse it to the wall of the restricting area. In time, the restriction will grow larger as the particles continue to collect. If the equipment is not maintained properly, the restriction will close off the path of travel. Figure 6.12 shows some hypothetical examples of impact fusion.

The Faraday cage effect

I have already mentioned the Faraday cage effect. Perhaps now I should explain it. Many people will tell you the Faraday cage effect is

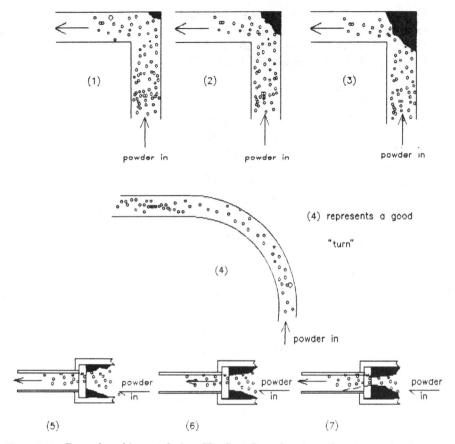

Figure 6.12 Examples of impact fusion. The first three drawings show increasing impact fusion at a 90° corner; the fourth shows how the situation can be avoided with better layout design. The last three drawings show increasing restriction from impact fusion at a passage point.

the "monster" of electrostatic spraying. I make it a point not to discuss the subject too much, because its effects are frequently overrated, exaggerated. Let's keep it simple. Let's say that the *Faraday cage effect* usually occurs when an attempt is made to coat an inside corner or the inside edge of a 90° bend in a metal part. The natural tendency is to aim the output of the gun straight into the corner, which results in an intensified ionizing field in the area. This, along with the air pressure the powder is being sprayed with, and the assisting air pressure, tends to set up both an electrical field and a strong air pressure in the corner you are trying to coat. This in turn repels the powder and prevents it from entering the corner. However, if the gun is aimed at the same inside edge from a different angle, the air and the ionizing field may be forced from the area as the powder is deposited, thus partially eliminating the problem, at least enough to let the powder and its charge cling to the metal part. Figure 6.13 shows some examples of where Faraday cages occur and how to avoid them.

Back ionization

Back ionization, or reverse ionization as it is sometimes called, is a result of attempting to apply too much powder to a particular part. As you apply the powder, the ground becomes "hidden," "lost," or insulated from powder particles. As the powder continually tries to land on a surface already coated, a charge builds up on the surface and eventually expels itself in the direction of the gun, which is grounded. It actually "spits" powder toward the gun and operator. The result is a craterlike surface, with heavy deposits of powder surrounding the cra-

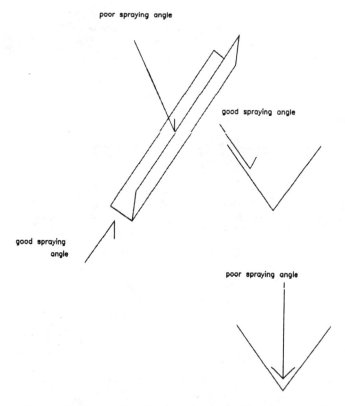

Figure 6.13 Overcoming the Faraday cage effect. Spraying directly into an inside corner can cause a Faraday cage; changing the angle of spray so that it comes from either the end or the edge allows the charged powder to cling to the part and gives the compressed air an avenue of escape.

ter. This is a rare phenomenon that varies from powder to powder and usually occurs only when an operator is first learning to spray manually. Experienced operators know when too much powder is being sprayed onto a surface, and they'll adjust their spray patterns or guns to reduce the powder flow.

Spraying Enclosures

Figure 6.14 shows some examples of the different types of spraying enclosures, which are basically three: (1) automatic, (2) manual, and (3) a combination, that is, an automatic enclosure with one or two manual touch-up stations.

Enclosures. Why "enclosures" and not the common term "booths"? My old dictionary defines "enclose" as "to shut or hem in." It defines "enclosure" as "the state of being enclosed." When you are powder-coating, you want to apply powder to the parts within the enclosure, not around the whole room. The enclosure "hems in" the overspray and removes it with the aid of the reclaim system. My dictionary does not define a booth as the place where I would want to spray powder. Powder coating is a class act, and if the system is set up properly, it will contain all powder not sprayed onto parts. During my years in the finishing industry, I just do not recall ever being around a wet finishing system which did not give off some aroma. The aroma was not shut off or hemmed in. It was not enclosed.

When you are planning your powder system, your first question will probably be "How many enclosures do I need?" Well, there are several ways to answer this. For example, suppose you're spraying 95 percent black, with just occasional touch-ups. Two separate enclosures would work well for you, one dedicated totally to black automatic and one for black touch-up and the other 5 percent miscellaneous colors you might occasionally coat. Figure 6.15 shows such a system.

Enclosures are made of steel, stainless steel, and dielectric plastic materials; they are sometimes painted and sometimes not. They come in all sizes and shapes. They usually have vestibules. Most have an opening in their roof so that conveyorized hanging fixtures can present parts to accept powder. The distance between the conveyor and the roof of the enclosure will vary. I like to see at least 24 inches between the bottom of the conveyor and the top of the tallest part. (See Fig. 6.16.) The idea behind this distance is to prevent powder dust from migrating onto the conveyor chain. Powder-coated conveyor chains, along with high-temperature lubricants, cause extra friction on the conveyor chain and in some cases also cause insulation of the chain from the hanging fixture.

Some enclosures are floor-mounted. Some are supplied with casters

(a)

(b)

Figure 6.14   Various types of powder enclosures. (*Courtesy of Nordson Corporation; Ransburg-Gema; Volstatic, Inc.; and Thorid Electrostatic Powder Systems*)

(c)

(d)

Figure 6.14   *(Continued)*

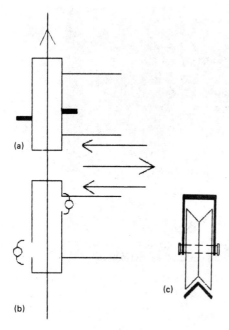

Figure 6.15 A two-color powder system. (*a*) An automatic unit that sprays 95 percent black and does all of the automatic coating; (*b*) a manual unit that handles all of the black touch-up and the remaining 5 percent of miscellaneous colors; (*c*) side view enlargement that shows how both units are mounted on tracks with caster wheels.

and tracks so they can be moved off-line to be cleaned. There are many methods for using such enclosures. The general concept is shown in Fig. 6.17, but many other methods are available on the open market. One such method is marketed by Thorid Electrostatic Powder Systems, and is called CHAMELEON. It is shown in Fig. 6.18.

Reclaim Systems

At the same time you are considering your spray enclosure, you'll have to also consider the various reclaim systems. In fact the enclosure and reclaim systems are often times talked about as one system—AERS, or the *application enclosure/reclaim* system.

These systems remove powder in several ways. Some manufacturers let the overspray powder travel the length of the enclosure, where it is drawn to an exit duct. It then departs via the duct work to the reclaim system. Figure 6.19 shows the general idea of this system. Some systems remove most overspray powder via a cartridge reclaim system located at the side of the enclosure. Figure 6.20 shows an example of this system at work.

Some systems let overspray powder go to two locations. The large heavy overspray particles fall by gravity to a fluid-bed feed hopper located in the center bottom of the enclosure. The lighter particles and

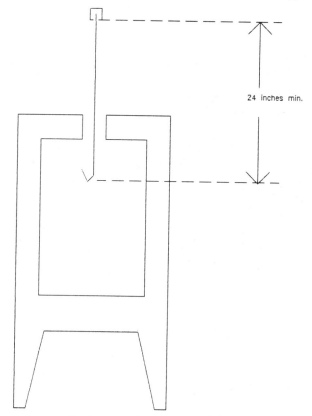

24 inches min.

Figure 6.16   Enclosure and conveyor. To help prevent contamination to the overhead conveyor system, there should
be a minimum of 24 inches between the conveyor and the
coating enclosure.

unusable dust are withdrawn to a reclaim system at the exit end of
the enclosure. Figure 6.21 shows this method.

Some enclosure/reclaim systems draw overspray to the floor of the
enclosure, to a porous moving belt which has a plenum under it. The
overspray is drawn to this carpet, which moves toward the exit of the
enclosure. When it reaches the exit, a vacuum head removes the
overspray powder to a reclaim system. The floating dust is removed by
a secondary vacuum system. See Fig. 6.22.

However it is organized, visualize a reclaim system as a very large
vacuum cleaner. Its size and shape may change, but its primary function is to remove oversprayed powder dust from the coating enclosure.
It does this by creating a negative air pressure within the enclosure.
Under the negative air pressure, the oversprayed powder and dust are
removed to a location to be temporarily stored and eventually reused

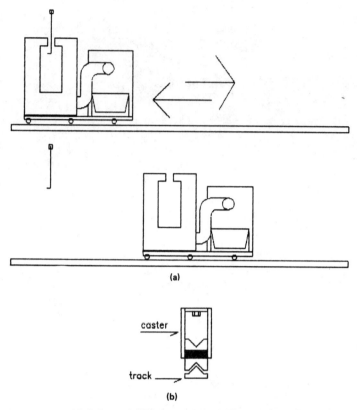

Figure 6.17 Modular on/off-line system. (*a*) The powder coating enclosure/cartridge system is mounted on a platform with (*b*) casters and tracks in line with the conveyor system. As needed, this unit can be rolled off-line to be cleaned and another unit rolled on-line in its place.

in the system. Figure 6.23 shows a simple vacuum cleaner and how it works. Some vacuum systems use cartridge filters.

The oversprayed powder itself is handled differently from the unusable powder dust. In most enclosures, gravity will cause the large powder particles to "fall" to the floor. Each manufacturer has developed a method for removing this oversprayed powder. In some enclosures it must be manually pushed or "squeegeed" into the reclaim system. In some enclosures most of it is removed automatically. In some enclosures it falls into a fluid-bed feed hopper to be resprayed.

The original reclaim system was called "the baghouse." It worked well, and there are still many in operation. The baghouses I have seen consist of a chamber filled with bags. This chamber is surrounded by a housing which has a negative pressure induced by a fan. The idea is to

Figure 6.18 The CHAMELEON. (*Courtesy of Thorid Electrostatic Powder Systems*)

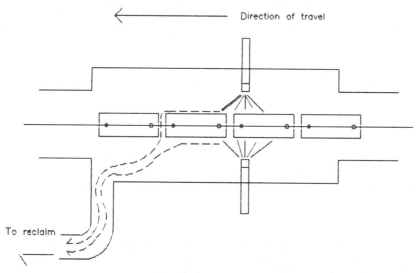

Figure 6.19 How a reclaim system works. The overspray powder follows the parts along the conveyor as it moves toward the reclaim system. The overspray particles, which still have a positive charge, have an "extra" opportunity to attach themselves to the parts. Here the design of the enclosure and reclaim system makes maximum use of the electrostatic phenomenon to increase the transfer efficiency of the powder to the part.

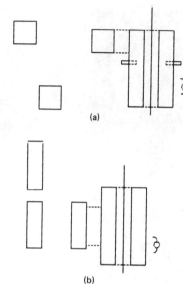

(a)

(b)

B

Figure 6.20 Two variations of a cartridge reclaim system. Each color has its own module. Whenever a color change is made in the system, the guns and enclosure are cleaned and the cartridge unit is removed and replaced. Having a dedicated cartridge for each color keeps both virgin and overspray powder all together in one module. Cartridge units are on casters and can be readily rolled to storage locations.

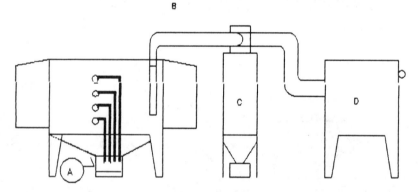

Figure 6.21 A reclaim system. The heavy oversprayed powder particles, by sheer gravity, fall back into the fluid-bed feed hopper (A); the lighter particles float out of the coating chamber and into the vacuum stream created by the reclaim system; they move through the system duct (B) to the high-efficiency cyclone (C); the usable particles remain in the cyclone, while the lighter particles of dust go into the cartridge unit (D) and the clean air goes back into the plant.

capture the overspray with the vacuum or suction and transport it to the baghouse where it collects on the interior walls of cloth bags, held there by the vacuum. Excess powder then drops into a container below the bags, to be stored and eventually reused.

The next generation of reclaim systems involved a cyclone and a final filter. This system was eventually replaced with a cyclone of

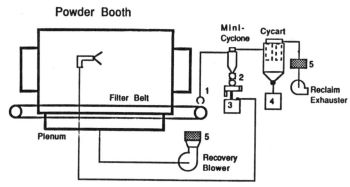

Figure 6.22 Multicolor filter belt recovery system: (1) the pickup head, (2) the pinch valve, (3) the hopper, (4) the scrap barrel, and (5) the final filter. (*Courtesy of Ransburg-Gema*)

higher efficiency and a final filter. Many people still feel this is the most economical method to use when many colors are being used.

At about the same time the high-efficiency cyclone was developed, the endless-belt system was introduced. This system has been improved many times and is still being used.

The cartridge system is a newer addition to the industry. It gives the user an opportunity to change colors easily by simply installing a different portable cartridge unit for each color. The cartridge system is very efficient, being self-purging and giving cleaner air back into your plant. It has received some notice because of this efficiency, and at least one company offers what it considers the best of both worlds: a high-efficiency cyclone, followed by a cartridge unit as a final filter. The reason for installing the cyclone prior to the cartridge system is to assist in quick color changing. Proponents of the system say that when a color change is involved, the only portion of the reclaim system that needs cleaning at the time of change is the container at the bottom of the cyclone, which captures usable overspray powder. The remainder of the powder fines which go into the cartridge unit are unused. Figure 6.24 shows how this might work.

AERS: The Whole Picture

Manufacturers are always interested in offering the people who use powder equipment the best system for the job. Some of the equipment companies offer more than one type of enclosure/reclaim system because they feel there are distinct differences in systems and by offering a variety, they can tailor their equipment to your particular work requirements. The illustrations at the end of this chapter will

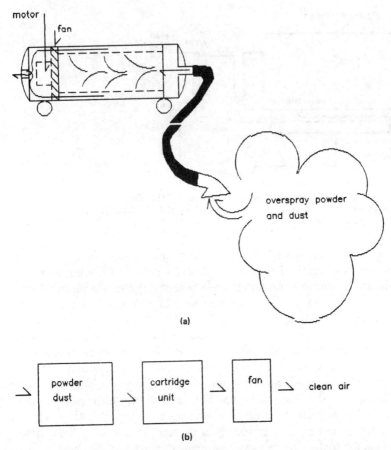

Figure 6.23 A simple vacuum cleaner system. The powder reclaim system works much like a vacuum cleaner. The dust and dirt are drawn by fan suction into the hose and then into the cartridge. The dust clings to the wall of the cartridge, while the air passes through the cartridge media. The now-clean air is blown by the fan and motor back into the room. Though powder system cartridge units are much more sophisticated, the principle is basically the same.

give you an idea of some of the equipment available from manufacturers.

Much of the talk you will hear when you are thinking of purchasing a powder system is of the pros and cons of one reclaim system or another. To begin at the beginning, let me tell you the main reasons for having a reclaim system. Quite simply, those reasons are (1) safety, (2) economics, and (3) cleanliness.

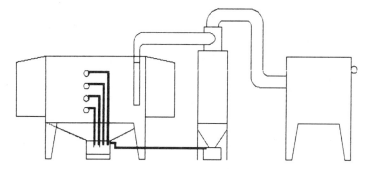

Figure 6.24 A high-efficiency cyclone/cartridge filter reclaim system. Proponents feel this system gives them the clean air of a cartridge system with the quick color change available with a high-velocity cyclone. The heavy particles in the overspray fall back into the feed hopper in the base of the enclosure. The lighter particles and the dust go to the high-efficiency cyclone, where the usable powder falls into the hopper in the cyclone base, and the dust goes to the cartridge unit. The powder in the cyclone base is then pumped back to the feed hopper. When a color change is needed, the enclosure, guns, feed hopper, and cyclone container must be cleaned.

Safety

It's a known fact that any dry organic substance that creates a dusty atmosphere when mixed with air will ignite when the proper concentrations of the substance and air meet a proper-sized spark. One could give many examples of this, but the simplest and most familiar is the example of common kitchen flour. By itself it is not dangerous, but when flour dust and air are concentrated in dense quantities, ignition can take place in the presence of a spark or flame. Sometimes static electricity can create the ignition. The grain industry has had many fires when grain is moved around in elevators.

It's to prevent this condition in a powder spraying enclosure that a reclaim system is used. It will remove powder overspray and dust from the coating enclosure via a suction device and put it into an area where the usable powder is separated from the air. The air is then forced through a filter system, so that the air that leaves the reclaim is clean. The air is now returned to the plant rather than exhausted to the exterior of the building.

Safety standards have been set up with regards to the ratios and concentrations of powder/air. Equipment suppliers can give you a copy of these standards. If you buy from a legitimate dealer/manufacturer of enclosure/reclaim systems, you can be fairly certain the proper calculations have been made, based on measurements of the openings of your enclosure, the quantities of guns being used in your system, and their potential powder output. Ask any manufacturer you are dealing with to specify the sizing of your reclaim system and ask to see their calculations.

A safety precaution often designed into automatic coating application enclosures is an ultraviolet (UV) ray detection system, which can detect electrostatic arcing. Of course, it would also detect flame. UV detection systems are integrated so that they shut down the powder system if they detect an arc. Arcs are usually generated when your housekeeping system is not working well. Hanging fixtures not properly grounded—that is, when cured powder has insulated them, for instance—can cause arcs between themselves, the parts, and the electrostatic guns.

If yours is a combination enclosure with some manual touch-up facility, you should have a UV detection system to cover those times when operators are not touching up the product and the system is running, unattended, automatically.

Economics

The economics of any material or process is important to the people who use the process, and to the eventual consumer. The cost you pay, or are willing to pay, for a product is based somewhat on its competitive pricing. As a result, manufacturers must heed this when designing, manufacturing, finishing, and marketing their products. Economic conditions available in any given enclosure/reclaim system work very simply, and they are based upon *transfer efficiency*.

People within the powder industry talk a lot about transfer efficiency. The manufacturers of powder talk about transfer efficiency. The equipment people talk about transfer efficiency. They'll all talk to you about the transfer efficiency of their products. The efficiency percentages they quote will usually run in "the high-90" range. I do not question these statements. But, achieving this efficiency requires help from the people working within the finishing area.

Let's take a close look at transfer efficiency. Figure 6.25 shows two hypothetical product lines to be powder-coated. One product line is composed of picture frames, the other of flat panels. Using the same powder and equipment, these two given product lines will get different transfer efficiency from both the powder and the equipment. Although the perimeters of both product lines are identical, the picture-frame line is full of holes. Through these holes passes much powder, some of which will cling to the walls of the enclosures. A powder particle hitting the wall might break up. If it does, it will increase the quantity of "fines," or minute particles which may not accept a powder charge.

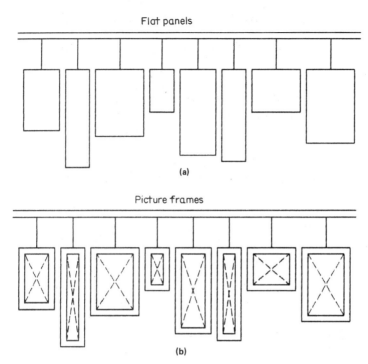

Figure 6.25 A case study of transfer efficiency. (*a*) Hanging flat panels as shown will establish a high transfer efficiency from the gun to the part. (*b*) The picture frames have the same perimeter as the flat panels, but the vacant spaces allow so much overspray that the transfer efficiency is drastically reduced.

This is not the fault of either the powder or the equipment manufacturer. The difference is within your product design and shape.

Operating your spraying equipment efficiently is important. Automatic guns constantly spraying the walls of an empty enclosure are making dust, not creating new highs in reclaim efficiency. A given pound of powder, in order to be 95 percent usable, needs to be sprayed onto parts, not onto enclosure walls and into reclaims. When not being used, automatic spray guns should be turned off.

When efficiency claims are made, they are usually made as a result of tests run on equipment in good working order, with equipment being used properly. A difference in 2 to 5 percent transfer efficiency can result when powder is spilled and sucked up into a vacuum cleaner or is sprayed on part hangers or conveyors. Buildup of impact fusion in remote parts of your equipment will reduce efficiency. So will many other points of day-to-day operation. But in spite of the many adverse conditions, with a little tender loving care, your efficiencies will remain very high.

Cleanliness

Powder dust, when not contained within a reclaim system, will float and migrate around your plant. When it finds its way in and around electric motors which run at a relatively warm temperature, the powder will cling and start to gel. If there is enough heat generated, the powder will eventually cure. Electric motors with cured powder coatings in the wrong places tend to become useless and eventually need to be replaced.

A clean finishing area eliminates contamination. I have seen many powder systems in plants all over the world. In some of them, you could spread a blanket on the floor and have a picnic lunch any time of the day. In others, what can I say? And it's in these latter plants, of course, where the people complain the most about contamination.

Additional Factors in Powder Application Equipment

Handling and conditioning of powder

Powder is sometimes shipped on request in fiberboard drums. If you use large quantities of one color, it's much easier to operate from one drum of 250+ pounds during a day rather than to open a lot of individual 45-pound boxes. Figure 6.26*a* shows how a drum can be handled. Figure 6.26*b* shows a drum-handling device.

Powder works best if it is "conditioned," that is, if the corrugated container holding the powder is opened and the plastic bag holding the powder is exposed to the general temperature of the area where the powder is to be used. This is especially important if the powder itself has been in storage in another area and there is a temperature differential. It is also important for the powder to be fluidized for some time prior to its use. This will permit it to flow easily when it becomes mixed with the compressed air and is pumped through the various hoses and equipment within your system.

Grounding of equipment

Grounding all of the application equipment is very important. As the powder travels through your system, it will pick up a slight static charge. By completely grounding your application system, you ensure

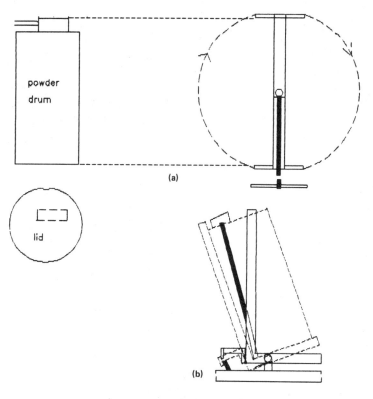

Figure 6.26 A powder drum. (a) The lid of the drum is removed and replaced
with a lid having a built-in powder pump; the drum is then placed on a stand
and rotated. (b) The lid is removed again and replaced with a new lid, which
also has a built-in powder pump; the drum is placed on a unit that is tipped at
an angle and locked in place. A vibrating device causes the powder to migrate
to the area where the pump can pick it up. (c) Photo of a drum-handling system. (*Courtesy of Ransburg-Gema*)

that these small charges will be eliminated. It is also important that you ground your conveyor hanging fixtures and parts as they pass through your powder application equipment. Without grounding, the transfer efficiency of your powder guns will rapidly go down to 0. Figure 6.27 shows a good method of installing a ground system in your plant.

Touching up

One question which always comes up when spraying methods are discussed: "Should I touch up before or after the automatic gun has done its job?" My answer to the question reflects my personal experience. I know that electrostatics work best when parts are well ground-

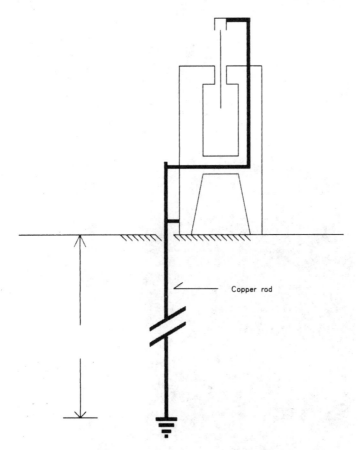

Copper rod

Figure 6.27 A simulated grounding system. The solid copper rod is driven into the ground to a depth of approximately 8 feet. Above the floor line, a copper ground strap or cable is attached from the rod to the enclosure; another strap is attached from the rod to the conveyor.

ed and clean. To touch up a part in a Faraday cage area, a deep recessed area, or any difficult area after the automatic guns have done their job means you have less clean or bare, grounded space to work with. Whereas a virgin part, with no powder coating, leaves much clean space for the touch-up of a manual gun. Then, when the automatic guns follow, the powder expelled will fill in all the bare, clean space, putting less powder in the vicinity already coated and giving a balanced coating thickness.

Your next question might be "How do I know where the Faraday cage areas or other difficult areas will be?" A good imagination will tell you. Then too, the first time you run production, your employees will see where the bare spots occur. They'll first attempt to fill these voids by readjusting the automatic guns. Then, as a last resort, they'll go to the manual guns to fill in the bare spots very quickly. It's merely a matter of familiarization and basic training.

Keeping records

Your system will work only as well as the records you keep. For instance, whether you are using automatic or manual guns, it is important to keep records of gun settings and positions. The guns work every day, but the people who control them take vacations and sick leave. Figure 6.28 shows a sample of the type of records needed for doing parts.

Making the Final Choices

It's been said that "No one AERS system will do the perfect job for every situation." A complete separate book on the subject might be able to cover all of the different systems for all of the different situations. But the time will come when your choices will have to be made. Make these choices in a systematic way. List all the requirements your company will actually need in an AERS. Then submit your list to the companies with whom you're negotiating for equipment.

Your greatest interest should be in how well your parts are powder-coated at the manufacturer's coating facility. Their sales engineer might be a very nice person, one you've grown to like and would like to retain as a friend after the sale is completed. But how well does that engineer's equipment perform during the test? Were you actually present when the tests were being run? Are you provided a videotape of the test to show your management as evidence?

If you are coating many parts, the demonstration should be performed with totally automatic equipment, if possible. You should have positive proof your parts can be done automatically. If touch-up is needed during

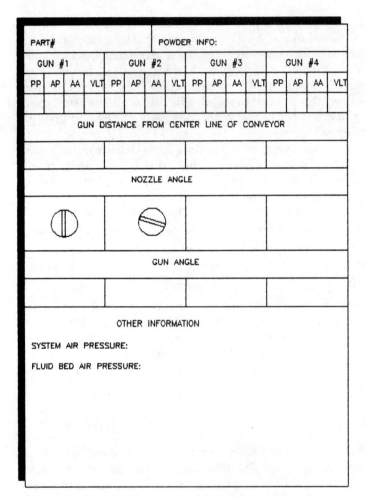

Figure 6.28 Chart for automatic gun settings.

the tests, then you know you'll need it in your system. If it isn't needed during the tests, you shouldn't need it during actual application, provided, of course, that your equipment is identical to the test equipment.

If, after the tests are run, you're happy with the results, you can start serious negotiation over your requirements for an AERS. Establish the answers to such questions as:

How many colors will I be using?

What percentage of my production will be allocated to each color I spray?

How often will I be changing colors?

Can the frequency with which I change colors be changed?

How long will it take me to change colors?

How long and how many people will it take to change colors?

How much will it cost me to change colors each time?

Do I need one enclosure or more?

Do I need one reclaim system or more?

How large an opening should my enclosure have?

How do I get rid of the small, unusable powder particles which float out of the enclosure and into the reclaim system?

Will the reclaimed powder be worth saving?

How much is my AERS going to cost?

Time equals money. With any portion of the manufacturing process, you necessarily calculate its cost and efficiency. It's no different in planning and purchasing a powder system.

The 30-minute system-cleaning time quoted to you for a powder coating enclosure with part openings 5 feet wide by 9 feet high and a total overall length of 35 feet (including the manual touch-up stations) was probably given to you by someone who indulges in the use of another type of powder, unless of course you have one of the systems specifically made for fast or quick color change. I have seen complex enclosures which take 2 to 8 hours to clean properly. Don't let anyone tell you that it's simply a matter of using enough people to do the job. How many people can clean a given enclosure/reclaim system? It isn't the quantity, it's the quality of the people and of the available cleaning equipment.

Let's talk about hypothetical situations. You have told your powder equipment sales engineer that you use four colors. You need to know how much equipment it will take to do the four colors. You could use from one to four enclosures and from one to four reclaim systems. It will depend upon:

1. How many colors you use each day

2. How many hours you spray each color you use on a given day

3. How much production time is needed for coating your product each day

4. What you do with the finishing line workers when you are changing colors

5. The opening size you need for your parts

6. Whether your spray system is automatic, manual, or a combination

The more equipment you need, of course, the more money it will cost. And what alternative will be open to you if your needs change in a year or so?

Each equipment manufacturer will have some good ideas for how to coat your product and exactly what equipment it will take to do the job. On one of our past project management jobs, we talked to three equipment people about a given project. The three companies each had a unique approach for the project. We permitted the companies to run their own tests, to assure themselves, and us, that they could in fact do the job the way they thought they could. In running the actual tests, we all found out that the parts were more difficult to coat then had been expected.

Coating your parts can be looked at from several different angles. Look at equipment being operated in other factories; talk to the people on the finishing line who clean and maintain the equipment. Tell them what you want to do on your line; get their ideas. People who actually operate and clean enclosure/reclaim systems will tell you exactly how they do it and how long it takes. They will also tell you how much their maintenance costs are and how often they have to replace parts.

Standard available application equipment

Many companies in North America manufacture powder coating application equipment. Some of these companies got their start in Europe. In fact, some equipment now sold in North America is imported from Europe and the Far East.

North America is a lively market, and much application equipment is sold here. In fact, there is so much equipment sold, the Powder Coating Institute has been keeping records as to the increases of sales in the automatic equipment area. The latest numbers available, along with the numbers for the past couple of years, will give you an idea of the leaps and bounds made within the industry.

I feel it is important for you to see some of the various pieces of spraying equipment available on the market. The following photos and drawings show samples of some of the standard equipment available to you and your company. By no means is this to say these are the only types of equipment available. Because of the time involved, I did not contact all of the equipment manufacturers in North America. For this, I apologize.

Figure 6.29 gives examples of some of the hand gun/control console/ feed hopper combinations available "off the shelf," so to speak.

Figure 6.30 shows examples of some of the enclosure/reclaim systems available. Equipment designed for your product could have part

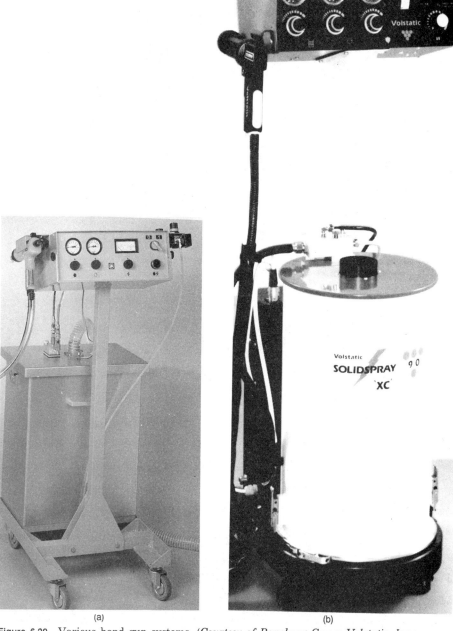

(a)   (b)

Figure 6.29   Various hand gun systems. (*Courtesy of Ransburg-Gema; Volstatic, Inc.; Nordson Corporation; Thorid Electrostatic Powder Systems*)

(c)   (d)

Figure 6.29   (*Continued*)

(a)

Figure 6.30   Various enclosure/reclaim systems. (*Courtesy of Nordson Corporation; Volstatic, Inc.; and Ransburg-Gema*)

(b)

(c)

Figure 6.30   (*Continued*)

openings of from 12-by-12 inches on up to sizes big enough to accommodate some large pieces of agricultural equipment.

Special application equipment

In an effort to accommodate the various industries we want to serve, the powder industry has many times gone out of its way to design, manufacture, and install a totally new piece of equipment designed to coat a specific product. Many long hours have gone into the development of specialized pieces of spraying equipment. I think it is important for you to see pictures of some of these spraying units. Who knows? Maybe the equipment you are looking for to coat your product or parts is one someone else has developed, with the research and development costs already paid for.

One piece of equipment available is called the electrostatic fluidized bed. With its assistance, many unusual jobs can be done. Figure 6.31 illustrates the basic electrostatic fluidized-bed principle. Variations of the electrostatic fluidized bed are used to coat what one might call endless-type products, such as wire cloth and fencing, on a machine similar to the one shown in Fig. 6.32. Other machines, like armature and stator coaters, handle various types of electrical components that require precise but heavy coatings. Figure 6.33 shows some typical

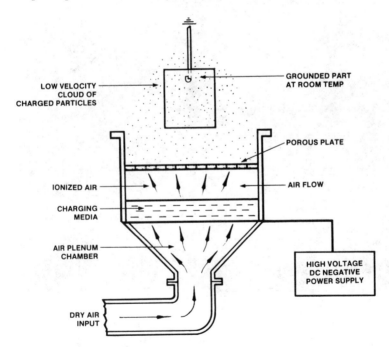

Figure 6.31 An electrostatic fluidized bed. (*Courtesy of Electrostatic Technology*)

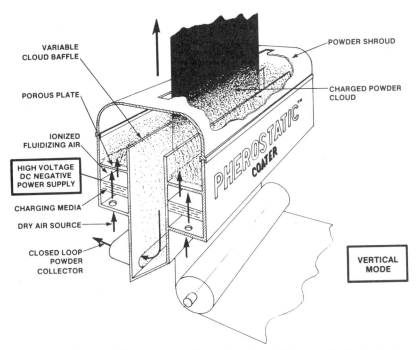

Figure 6.32 A wire cloth coating machine. (*Courtesy of Electrostatic Technology*)

Figure 6.33 Typical electrical components coated in an electrostatic fluidized bed. (*Courtesy of Electrostatic Technology*)

electrical components, and Fig. 6.34 illustrates an automatic electrostatic fluidized-bed system that coats and cures armatures.

The application of powder to a mold cavity is another specialized process. The actual coating application occurs while the mold halves

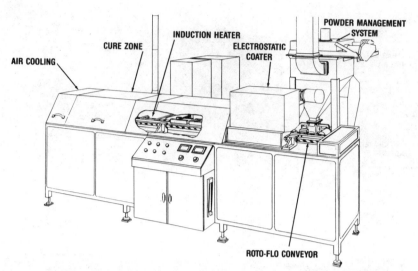

Figure 6.34   System of electrostatically coat and cure motor armatures. (*Courtesy of Electrostatic Technology*)

are separated. The powder is applied to the heated cavity; the plastic substrate is then inserted into the cavity. The heated cavities come together and the partially cured powder is transferred to the plastic substrate where the cure is completed. The process, known as in-mold coating, is fairly new in North America, but there are at least two companies using it, and it appears that this will be an interesting market for the powder industry. Figure 6.35a shows a cross section of such a system, and Fig. 6.35b shows a simulated system at work.

A number of companies have put together chain-on-edge, or vertical-mounted spindle conveyors. Occasionally a special type of project comes along which lends itself to this type of application. Figure 6.36 shows a photo of one of these systems in operation.

Some powder-coating application systems are especially designed to change colors; they use a variety of methods. Figure 6.37a and b shows the ColorSPEEDER, a system specifically made to change colors quickly. The ColorSPEEDER has been shown at several industry trade shows. Another specialized machine is the FREEDOMCOATER, which is an automatic powder coating machine (Fig. 6.38). It can coat one side of a product such as brake shoes, brake drums, and flat sheet without masking.

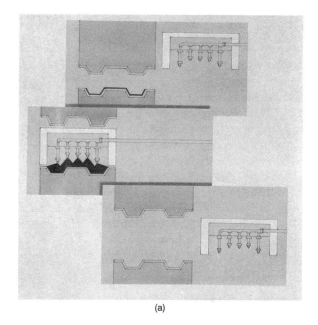

(a)

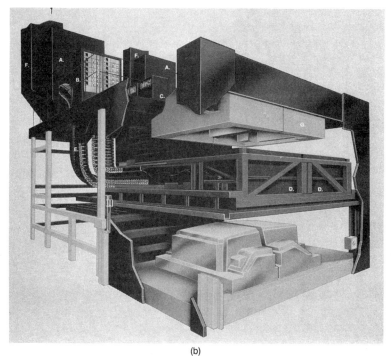

(b)

Figure 6.35 (a) In-mold coating process. (b) In-mold machine: (A) cartridge-filter collector; (B) spray gun control consoles; (C) feed hoppers; (D) coater boxes with spray guns; (E) exhaust duct; (F) pressure vent. (*Courtesy of Nordson Corporation*)

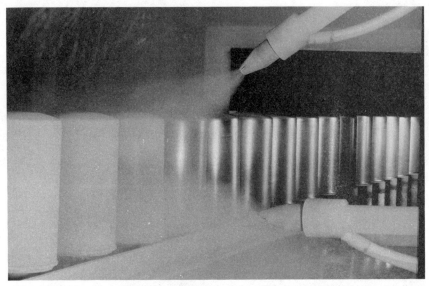

Figure 6.36   A chain-on-edge powder spraying system. (*Courtesy of Nordson Corporation*)

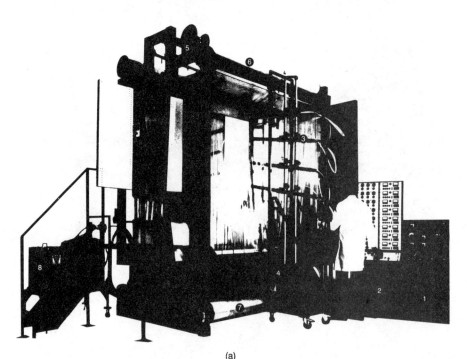

(a)

Figure 6.37   The ColorSPEEDER. (*Courtesy of Volstatic, Inc.*)

(b)

Figure 6.37 (*Continued*)

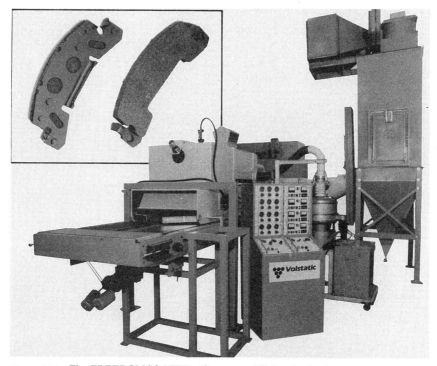

Figure 6.38   The FREEDOMCOATER. (*Courtesy of Volstatic, Inc.*)

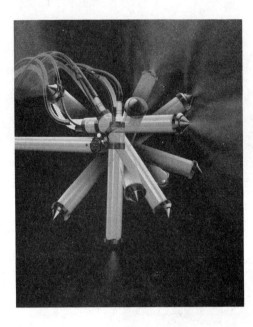

Figure 6.39 A swivel-mounted automatic spray gun. (*Courtesy of Volstatic, Inc.*)

Figure 6.39 shows a swivel-mounted automatic spray gun, designed to reach almost anything difficult to reach and coat; the availability of such a piece shows the dedication of the equipment manufacturers to the finishing industry.

7

Ovens

Overview

The oven is a very important part of a finishing system. Most ovens are large and consume a lot of expensive floor space. Although insulated, ovens also radiate heat, which can become objectionable to employees and can adversely affect the temperature of cooling parts. However, once an oven is installed and balanced properly, it will give you less problems than most of the equipment you have in your plant. A bit of tender, loving care is all an oven needs to give you many years of service.

Ovens are used in several areas of powder coating. They are used

1. As dry-off ovens after parts have completed the pretreatment cycle

2. As cure ovens after the powder coating has been applied

3. As preheat ovens for parts normally finished with thermoplastic materials and, if necessary, for the postcure after the application of the materials

4. As burn-off ovens to remove unwanted cured coatings on hanging fixtures and/or reject parts.

Oven Fuels

The three common fuels used in ovens within the finishing industry are natural gas, propane gas, and electricity. *Natural gas* is the most economical and available, and thus the most popular, fuel. *Propane* ranks second in this regard, followed by *electricity*. In most plants you'll find natural gas or propane, but there are plants that must rely on electricity simply because it's available and it's cheap, and it can be used in both infrared and convection ovens.

In many states it is mandatory to set up dual fuel systems for your equipment. Normally, natural gas is the primary fuel, with propane used as a standby fuel source. During the winter, the natural gas company may frequently telephone your company and advise you that it's necessary to switch over from natural gas to propane, usually giving you an hour or two to make the change. Your maintenance people then open or close a series of valves, and your system is switched to propane without a shutdown.

Types of Ovens

The basic types of ovens used to apply heat to parts are *convective ovens* and *infrared ovens*.

Convective ovens

Convective heat is the most popular type of heat used within the industry. A convective oven convects, or transfers, heat from the oven heating system through the use of hot air, which is generated in an area alternatively called a firebox, heater box, or any one of several other names. From this chamber, through a set of heat ducts and a recirculating fan system, the hot air is directed to the parts to be cured. The air is then returned to the heating chamber, reheated, and the process repeats itself. A portion of the returning air is exhausted, and fresh ambient air is introduced to the burner. Figure 7. 1 shows

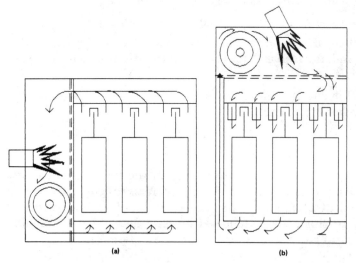

(a)   (b)

Figure 7.1 Two convective oven systems: (*a*) Oven with a side-mounted burner that recirculates hot air from the floor to the ceiling. (*b*) Oven with a burner mounted on its roof that sends hot air down on parts; the return air cycles at the floor of the unit.

two different convective oven systems. Designs vary from manufacturer to manufacturer.

Infrared (I/R) ovens

Infrared heat is created by using invisible infrared rays from the low end of the infrared light system; parts are saturated with direct infrared rays, causing them to rise rapidly in temperature and cure very quickly. Two types of fuel—gas and electricity—are used in heating infrared ovens. Because of their speed and efficiency, infrared ovens are much smaller in size than normal convective ovens. Limited floor space alone can make infrared ovens the optimum choice.

Infrared ovens have certainly found a home in powder-coating systems. In many instances they have been used not only to complete a cure but to solve other problems as well—such problems as arise when the great weight of your parts or the quantities to be cured would normally call for an oven size that could overwhelm a fast-moving production line.

I/R ovens use infrared in the high-, medium-, and low-density ranges. In deciding on an infrared oven you must take into consideration the part to be cured. In my experience, infrared ovens can cure parts with some complexity to their shape relatively well, but as the shape becomes more complex, a tradeoff point is reached and you have to turn to a convection oven.

The high-density electric infrared oven serves the industry well in a few special areas. For instance, it can cure special epoxies and specially formulated powders within a few seconds.

The low-density electric infrared is used as a cure oven, having the ability to cure certain films in a time span of between 3 and 6 minutes.

Without getting too technical, I would like to touch on some comments you'll hear regarding low-intensity infrared ovens. Opponents of the I/R system will say, "If your parts are flat sheets, I/R is good, but it can cure only what it sees." Proponents claim that that is not completely true, saying "The interior of I/R ovens is usually a polished surface, and stray I/R rays will bounce from these reflected surfaces onto the product." Also, parts accepting I/R heat conduct it rapidly, they claim, and quite quickly all the surfaces of the part are heated.

Types of electric infrared generators vary with manufacturer. Figures 7.2 and 7.3 show typical examples of infrared ovens; they also show the infrared generators themselves, attached within the center of a highly reflective cone-shaped surface.

A typical gas infrared oven incorporates a number of specially designed gas burners built into the walls of the infrared section. These

Figure 7.2   An infrared oven. (*Courtesy of Dry Clime*)

Figure 7.3   An infrared oven. (*Courtesy of Dry Clime*)

burners are generally built into stainless steel reflectors, which reflect as much heat as possible to the parts being cured.

Let's look at some examples of the advantages gained in using an infrared oven. Suppose the normal cure cycle for a given part is stated at 15 minutes, 360°F in a standard convection oven. At a line speed of 5 fpm, it would take a minimum of 75 feet of oven time for curing, plus some time allocated to heating the part up to temperature. Yet a low-density infrared oven could heat and cure the part in from 3 1/2 to 6 minutes. Figure 7.4 shows how this could work.

Sometimes gas infrared is used as a booster for preheating parts prior to their entering a convective cure oven. In this case, the infra-red brings the part up to temperature usually within a minute or a minute and a half, thus reducing the overall heat-up time and size of a cure oven. Booster infrared ovens can be used when converting from a wet-system, low-temperature coating to powder; when the product is heavy and a long heat-up time is required, but space is limited; or when production rates have escalated beyond the capacity of the oven in use. Figure 7.5 shows how this might work.

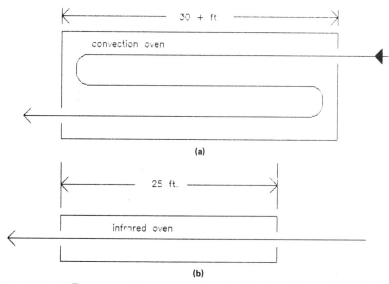

Figure 7.4 (a) For a convective oven, a cure cycle at a given line speed of 5 fpm is 15 minutes, which would mean a required footage of 75 feet. (b) For an infrared oven using average time and temperature, 25 feet is sufficient. Not only is the floor space required considerably smaller, but because of its light weight, the I/R oven might even be ceiling-mounted.

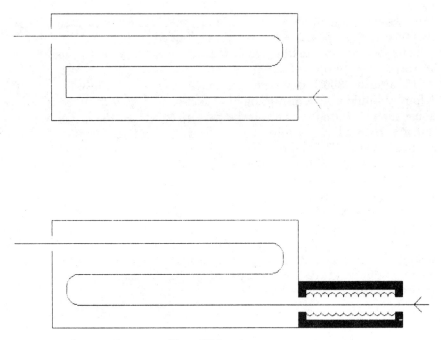

Figure 7.5 A convection oven with an I/R booster.

Combination ovens

Combination ovens can be set up to accommodate a dry-off cycle and a cure cycle within the same oven. Depending upon the manufacturer and the design used these are made in several different ways:

- *Two ovens, two burners, and two recirculating systems, with a common wall between the two oven sections:* Though this system increases the costs over the single oven, single burner system, it is sometimes necessary to do the job. Figure 7.6 shows a sample of this type of setup.

- *A single oven and single recirculating system, with a common wall between the two systems and with dampers to control the temperature in the dry-off section:* Figure 7.7 shows an example of this setup.

- *A single oven with no wall between the two processes, where the governing factor is the time needed for the part in the dry-off section:* As an example, Figure 7.8 illustrates the case where the dry-off time would be approximately 5 minutes, and the cure cycle would be 15 minutes: one pass for the dry-off cycle and three passes for the cure cycle. The dry-off cycle is adjusted by the length of the oven.

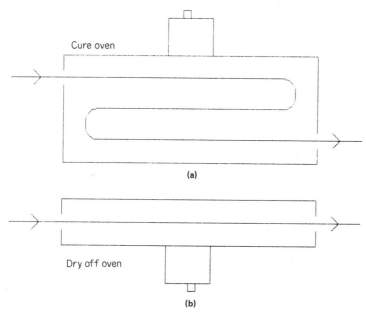

Figure 7.6   A combination dry-off/cure oven: (a) a self-contained, three-pass cure oven and (b) a one-pass dry-off oven. Each oven has its own burner and recirculating system.

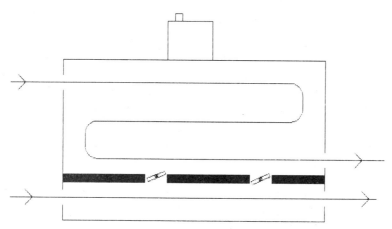

Figure 7.7   A combination dry-off/cure oven. Both ovens are within a single housing. They both use the same burner and recirculating system. The dry-off and cure sections are separated by a wall; dampers, along with the dwell time, control the temperature of the parts in the dry-off section.

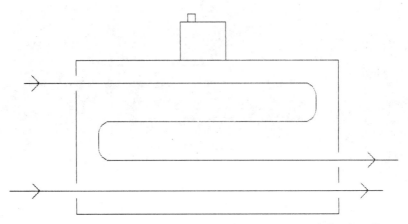

Figure 7.8 A combination dry-off/cure oven. The cure oven and the dry-off are within the same housing. The cure takes three passes, and the dry-off takes one pass. There is no wall between the two processes. The dry-off time is approximate and is based on the mass of the product being dried.

Hot-Air Oven Recirculation Systems

Each oven designer has his or her own idea of how to heat and design recirculation systems to move the hot air within an oven. It's standard for the air to be heated and circulated through a ducting system. A typical system is shown in Fig. 7.9. The hot air passes over the parts, gradually raising their temperature. The air must then be drawn away, cycled back into the firebox, reheated, and again blown over the parts. During this endless cycle, a portion of the air must be removed from the oven and replaced with fresh, clean, ambient air, which is then raised to the temperature of the firebox or combustion chamber.

Firebox locations and designs change according to the total overall oven design. Some designers bring in the hot air through ducts in the floor of the oven, letting it exit at louvers above the parts or through other return-air duct systems, which take the hot air, now reduced in temperature after passing over the parts, back to the firebox for reheating.

Again, you must remember each oven manufacturer has its own basic design for making the air move properly around your particular parts. The important thing is that there be a large number of adjustable louvers through which the air can pass on its way to the parts, as well as a channel by which the air can be removed from the oven for reheating. The louvers are what makes it possible to adjust, or balance, your oven as the parts pass through it on a conveyor system. The amount of air entering at each foot of the conveyor path within the oven can be regulated very carefully. A large number of louvers will keep all of the air circulating and prevent the accumulation of dor-

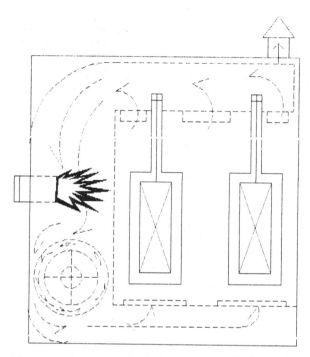

Figure 7.9 An oven recirculation system, which heats the air within the oven by forcing it to recirculate past the gas burner. When the air is heated, the recirculating fan forces it into a duct system, which exits into the oven through adjustable louvers in the floor of the oven. The hot air rapidly rises and is drawn through the return-air louvers, reentering the recirculating system. A portion of the air is exhausted from the oven, and fresh air is brought in at the burner.

mant or nonrecirculating air at any given point. These points of dormant or slow-moving air, sometimes called "hot spots," can create havoc with the temperature control in the oven. They can cause discoloration in parts and they can overcure parts. They can also damage your conveyor system.

We have shown here only one design; there are other types of hot-air recirculation systems available on the market.

Oven Temperature/Part Temperature

Your oven is used to cure your parts. To cure them properly, the powder manufacturer will recommend a minimum cure cycle, for example, 15 minutes at 375°F. Adjusting the "set point," or temperature gauge, on your oven to 375°F does not guarantee your part will get the proper cure, since the reading on the instrument panel gauge only gives the oven temperature at the location where the probe is placed within the oven. It says nothing about the temperature of the part. You cannot, and must not, assume that your parts will reach the proper temperature within the oven, unless the oven is completely balanced.

Oven technicians usually balance oven temperatures by adjusting louvers located throughout the oven in the ducting system. The simple, thorough, reliable system I use for testing the balance is easy to follow:

1. Note the set point reading.

2. Fill the heated oven with your parts; set them on the racks you will be using when in production.

3. Solidly attach a thermocouple to a part at the top of your hanging fixture and run it through the oven, noting the temperature each minute for the length of time the part is in the oven.

4. Repeat the process, again attaching the thermocouple to a part, this time somewhere at the middle of the fixture, and again noting the temperature each minute for the length of time the part is in the oven.

5. Repeat the process a third time, this time attaching the thermocouple to the lowest part on the hanging fixture, and note the temperature each minute for the length of time the part is in the oven.

(Some temperature-testing units are equipped with multiple temperature probes, in which case a single test could be made with multiple probes attached to various locations of your parts and a graph reading taken for each probe.)

The results of this test will show you if your oven needs closer balancing. If you have parts with great weight variations, it might be advisable to run light parts during a separate test and then test your heaviest parts separately. This testing will take some time, but it will give you part temperatures at different elevations and at different points within your oven. It is most important that the parts all pass the minimum temperature and cycle times recommended by your powder supplier, or you will be wasting time and energy in the cure oven.

By getting a temperature reading every minute, you will be able to tell exactly where adjustments need to be made within your oven. As an example, given a line speed of your conveyor as 4 fpm, you can gauge your oven at

4 fpm × 1 min, or a point 4 feet from the entrance of the oven

4 fpm × 2 min, or a point 8 feet from the entrance of the oven

4 fpm × 3 min, or a point 12 feet from the entrance of the oven

Any way the oven testing is done, it is important that it be done properly and that a record or chart be made of the temperatures at each foot of travel within the oven. Figure 7.10 shows some of the information you'll need and gives some suggestions as to how you might note the approximate temperatures at certain points within the oven.

Some oven-testing equipment is completely self-contained, and

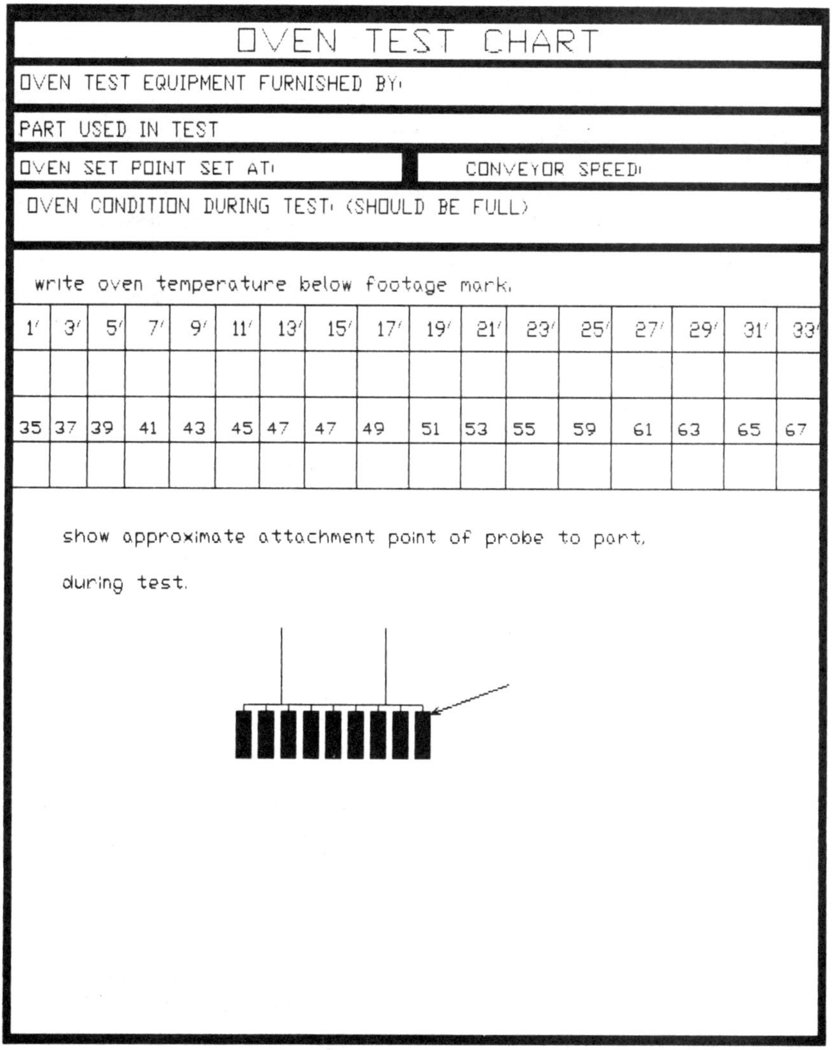
OVEN TEST CHART

OVEN TEST EQUIPMENT FURNISHED BY:

PART USED IN TEST

OVEN SET POINT SET AT: CONVEYOR SPEED:

OVEN CONDITION DURING TEST: (SHOULD BE FULL)

write oven temperature below footage mark.

1′	3′	5′	7′	9′	11′	13′	15′	17′	19′	21′	23′	25′	27′	29′	31′	33′
35	37	39	41	43	45	47	47	49	51	53	55	59	61	63	65	67

show approximate attachment point of probe to part, during test.

Figure 7.10 Example of a chart to record oven testing.

some is so sophisticated as to make complete graphs of temperatures as the parts go through the oven. Some testing equipment requires a cable attached to the hanging fixture, with the readout given outside the oven. If the attached cable is marked in 1-foot increments, the exact location of a louver adjustment to either increase or reduce the temperature can be noted.

Oven Location

When you are in the planning stage of your system, you'll have a choice of oven location. However, once the oven is installed, it will be difficult to move, so get it in the right location the first time. There are a variety of locations you can consider in the planning stage:

1. You can compactly fit the system all together on one level, being careful to see that the radiating heat does not interfere with any other operation in the plant.

2. You can suspend the oven from the ceiling, using supporting steel and exhausting the excess heat through vents in the roof.

3. You can place the oven on the roof of your plant.

4. You can mount the oven on a concrete pad outside of your building.

There are many factors to take into consideration in choosing among these options. It's great to locate your system compactly together, provided that you don't put the oven next to the powder booth. Radiating heat eventually will take its toll, and cure powder dust. If your oven air intake is close to the powder spraying enclosure, it may draw stray powder particles into the recirculation system of your oven. If you use more than one color, this could give you salt-and-pepper parts.

Suspending the oven from the ceiling gets it off your plant floor and out of the way, and permits you to use a bottom entry and exit, which helps keep the heat where it belongs, inside the oven. If you consider ceiling mounting, you'll have to also consider adding some steel for the support of the oven.

Roof mounting of ovens gets the oven completely out of the building. It also takes care of any radiating heat, which, although welcome in the winter, can be very uncomfortable in the summer. It is usually necessary (especially in the northern states) to box the oven in a small house for protection against the weather and for added insulation. It will also be necessary to add support steel to the roof in the area where the oven will sit. Depending upon the roofing material, it may be necessary to place the oven on rails or I beams, so air can circulate between the oven and roof and prevent the melting of roofing materials.

With some intelligent planning, an exterior oven housing can become a thing of beauty outside your plant. It is possible to house the oven with the same material used on the fascia of your building. With landscaping, the oven area can blend in perfectly with the surrounding area.

Ovens as a source of plant heat

One of the most interesting projects we were ever involved in was a project where we enclosed the power washer and the dry-off/cure oven within a room in the corner of a new plant in one of the northern states. We designed the conveyor to keep cured parts within this room until they cooled down, so heat radiating from the cured parts would remain within the room. Exhausts from all equipment exited the building as usual. We were only interested in controlling the clean heat, or the heat radiating from the parts and equipment. We shunted this heat into a series of ducts to heat the building or, if not needed for that purpose, to leave the building. The system was completely automatic, dependable, and kept factory employees warm in the winter. Figure 7.11 shows how a system like this could work.

Stacking-Type Ovens

When the system has to accommodate parts that are extremely long, fast-moving conveyor lines, or unusually large quantities of small parts, consider installing a stacking-type oven.

Suppose you have large quantities of 20-foot-long aluminum extrusions; these are to be hung horizontally for finishing. Because of their length, these parts would need large turns in the cure oven. Indeed, a very large curing oven would be needed for the 20-minute cure, plus the 5-minute cycle for the parts to come up to temperature. Suppose 1000 pieces per hour are to be cured and the extrusions are hung 28 per fixture, with 1 foot of space between fixtures. That means you'll need thirty-seven sets of hanging fixtures (1000/28) at 21 feet per fixture per hour. And you'll need a 105 × 50-foot oven.

Figure 7.12 shows how a normal oven might look versus a stacking-type oven in this situation.

Programmable Controllers

It is possible to start and stop your oven automatically with programmable controllers and/or other timing devices. Discuss this with your

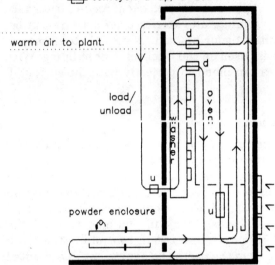

Figure 7.11   A heat-conservation system. The washer and dry-off/cure oven are located in a room which has a slight negative air pressure. Clean radiant heat from the exterior walls of the equipment and product rises to the ceiling of the room. If hot air is needed in the plant, it is drawn through the duct. Fresh air is brought into the plant through air-intake units which operate independently. In the summer, ceiling fans remove hot air as required.

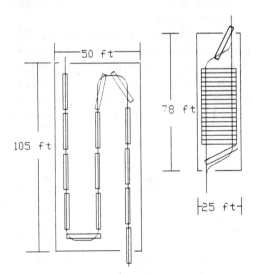

Figure 7.12   Comparison of a normal oven as loaded and a stacking-type oven.

oven vendor and consider the possibilities while planning your system. Perhaps there is a need for such a controller somewhere else within your finishing system. If installed properly, a controller can enhance the complete finishing system; installed improperly, it can quickly become a case of the tail wagging the dog.

Oven Exhaust

You have to exhaust your oven to the outside of your building. You'll probably hear that powder exhausts are not harmful, but, remember, the products of combustion coming from your oven will have to be safely exhausted. Federal laws govern the minimum amount of exhaust required for ovens. Your oven manufacturer is aware of these laws and designs exhaust systems accordingly. Generally your oven exhaust will be regulated through a damper.

Air Seals

Your oven should be equipped with a good set of air seals, which are designed to keep most of the heat in your oven, even though conveyorized parts cycle through it constantly. The majority of the heat felt at the openings is radiating from the oven interior and the cured parts. Good air seals will save you a lot of money in fuel costs by keeping the heat within the oven. Factory people usually adjust the seals according to the prevailing conditions in your plant. If at a later date, a new door creates a problem with the balance of the air seals, they may require readjusting. With ceiling- or roof-mounted ovens, which use a bottom entry and exit, openings do not require seals because of the natural function of heat rising.

Oven Lighting

Another important factor in oven design is lighting. When maintenance is required, or for any other reason you need to enter your oven, it won't hurt to have something like a dock light with a swinging arm hung from the outside of the oven for easy illumination of the oven interior.

Oven Specifications

Your oven manufacturer will probably suggest the proper path for your conveyor to follow through the oven. The manufacturer will want to make sure that:

1. Your parts will clear all internal framing and ducts; this is to protect the parts and conveyor from any blasts of flame from the oven burner.

2. Your conveyor is protected from the same ill treatment your parts are protected from, such as direct flame or excessive heat from the oven burner chamber.

3. Your conveyor is electrically tied to your oven, so that the conveyor is always moving while your oven is on; this will keep the wheel bearings and lubricants of the conveyor from becoming "baked."

4. Oven entrance and exit air seals closely control air currents, since strong currents can blow powder off parts and create a batch of rejects.

Specific figures you'll need to supply the oven manufacturers include:

1. Pounds per hour of parts, hanging fixtures, and conveyor chain going through your oven

2. The envelope, or overall dimension of your parts and the hanging fixture, so the proper-size part openings can be designed for the oven

3. Length of the longest part, parts, and/or hanging fixtures, so that when the package makes a turn, or in some cases merely makes a straight pass through the oven, parts and/or hanging fixtures will not touch the walls, which may cause powder to fall off the part before it has started to flow or, worse yet, may scrape partially cured powder from a part; rubbing against walls is also tough on your conveyor system

4. The exact cure cycle for your powder, that is, x minutes at x degrees (Remember, this is after bringing the part up to its cure temperature; you'll also need to allow for heat-up time within the oven before the part begins to cure.)

Oven Insurance

While you're still in the design stages, talk to your insurance carriers and your oven vendors about insurance requirements. Your insurance carrier can explain insurance requirements to you, and your oven manufacturer can then give you the true cost for the system, without having to change the price at a later date because of insurance requirements that had not been factored in earlier.

Oven Installation and Maintenance

When your oven is erected and ready to fire up, the company from whom you purchased it will probably start it up and adjust it for you. If they know what they are doing, they will balance it to accommodate your part loads. Make certain the oven performs properly before the oven company representative leaves your plant. I have no quarrel with oven company people, but once they leave, they are likely to charge you for a return trip to make adjustments.

Your oven is not just a great big chunk of iron. It is a finely manufactured tool. For proper operation, it will need occasional maintenance. It will need to be treated like any other piece of expensive equipment or it will fail.

The oven must be started in early morning so that there is enough time for it to heat to operating temperature. By the same token, it will be necessary to let your oven cool down before shutting it off for the day. The oven manufacturer will likely instruct your employees to reduce its temperature to somewhere about 150 or 200°F before shutting the recirculating system off, since shutting the oven off at temperatures higher than this could damage or warp internal parts of the recirculating system, causing much down time and costly repairs.

Oven Documents and Maintenance Records

It is important for you to keep some limited maintenance and historical documents for equipment. One of the things you should keep, as a matter of personal insurance, is a record of the heat curve of your oven as your parts go through it. By occasional testing, you'll be able to determine if your oven is keeping pace with production, or if it's time for some maintenance or remodeling.

8

Conveyor Systems

Overview

Most powder-coating finishing systems are production lines. When a company wants to produce large quantities of items that require a finishing process, a conveyor system is needed. Such a system will control the flow of parts through the entire finishing process, once you establish the path. A conveyor system can be equated with the sheepdog used by a sheepherder: The main function of a continuous-chain conveyor within a finishing department is to do the job with very little direction once it is designed and installed—and once the finishing-line personnel know what is expected of them.

A good, efficiently operating conveyor system will cost you less money than the variety of other material-handling devices you could use to move your product or part through your plant. A conveyor system that takes your parts from the point of manufacture, through the finishing department, and on to final assembly will eliminate many of your daily manufacturing costs.

For all intents and purposes, this chapter will discuss only the overhead monorail conveyor, though we will talk of both the continuous-chain variety and the popular power-and-free conveyor type. We will cover the moving of your product from the manufacturing area, through the entire finishing process, and on to the final assembly or packaging area.

Designing a Conveyor System

To start with, it takes a lot of experience and knowledge to design a conveyor system properly. There are many good conveyor engineers in the field; they have the knowledge needed to put a good system into your plant. If you want to do the best job you can for your company, arm yourself with the basic information as to what you want your new conveyor system to do when it is installed and operating, and then contact a good conveyor specialist. We'll give you some direction here as to the information you'll need when you first meet with the conveyor specialist.

Parameters to be considered in the design

Let's begin by listing the statistics and logistics to be factored into the design of a good conveyor system:

1. The weight of the individual parts (the heaviest and the lightest) you'll be hanging on the line.

2. The number of parts per hanger, if your hangers hold multiple parts, the total weight of these parts plus the hanger, and how the parts will be hung or grouped.

3. The physical dimensions of the parts (the largest and the smallest).

4. The amount of production you'll want to run through the system per hour, which will help determine the line speed. (You should plan for future expansion of your plant when you set up your hourly production rates.)

5. The conditions under which the conveyor will operate:
   The conveyor must be protected from damage from the water and the chemical solutions in the washer.
   The conveyor must withstand the high temperatures of the oven; a well-designed oven will prevent blasts of hot air from flowing directly onto the track and chain; the maximum temperature to which you should expose chain is 450°F.
   The conveyor must not be contaminated with powder as the hanging fixtures pass through the powder-coating enclosure.

6. Loading procedures in the exchange area, which is where you will stage or load parts for the finishing system and which is located at either the point of manufacture or at the beginning of the finishing system. Perhaps you will unload the finished parts at the exit of the finishing portion of the system, or perhaps they will be transported on to the final assembly or packaging area. In any event, can the parts be loaded on a continuously moving conveyor or must the carrier be stopped to be loaded?

7. Elevation changes, from the highest elevation needed in the conveyor system to the lowest. Will the parts clear the floor at the low-

est elevation? If you are using a power/free conveyor system, can the elevated areas be used for storage or only for transportation?

8. Safety standards set by OSHA and other safety guidelines. The part and hanger *envelope,* or *profile,* as it is sometimes called, is that area of space taken up by the width of your hanging fixture, including your widest part; the length of your hanging fixture and parts in the direction of conveyor travel; and the height of your package from the top of the conveyor itself to the bottom of the lowest hanging part. A safety factor should be added to each dimension. Figure 8.1 gives examples of envelopes. The upper section of the end view shows the area sometimes called the *keyhole.* Somewhere between 3 and 6 inches should remain between your actual part profile and the walls of your equipment. The upper area around the conveyor and fixture is usually protected while the conveyor is in the power washer in order to keep chemicals from spraying into the keyhole area. Two methods used are shown in Fig. 8.2a and b. One method is to shroud the conveyor opening at the bottom with neoprene or a similar material; the other is to shroud the entire conveyor track, then supply a volume of pressurized air sufficient to create a positive pressure within the track area.

Figuring the process times

To estimate process times within various pieces of equipment, it will be necessary to establish your system's hourly production rate. You'll have to establish some process times, or residence times, for each portion of the finishing process. Cool-down times between actual process times must be allowed for as well. You just can't expect employees to unload 300°F parts 10 feet from the oven exit because you forgot to allow enough conveyor in the system. *All* times from the loading/staging area to the unloading area should be closely estimated.

Use the simple layout shown in Table 8.1 as you start your system design. This process-time chart lists various traveling times and residence times, enabling you to estimate equipment size and total length of the conveyor system. The time between each process is shown as OT, or "outside time." (This chart assumes a continuous-chain conveyor system.)

Sometimes a block diagram is useful in visualizing a finishing-system operation; see Fig. 8.3. Whether you refer to Table 8.1 or Fig. 8.3, you'll get a good idea of the process feet in the total system and of the time spent in each process area. You can certainly use similar tables and/or diagrams if you are trying to "shoehorn" a system into a particular area of your plant or retrofit a powder system into your

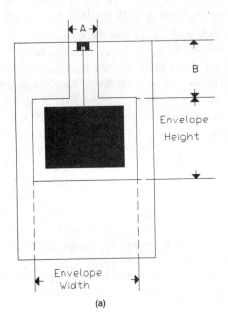

(a)

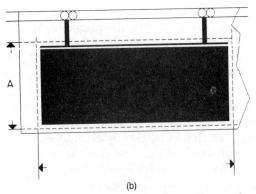

(b)

Figure 8.1 Examples of envelopes: (*a*) The envelope width and height are the part width and height, plus 3 to 6 inches on each side. A and B areas represent the keyhole. (*b*) The envelope length and height (A) are the part length and height, plus 3 to 6 inches on each side.

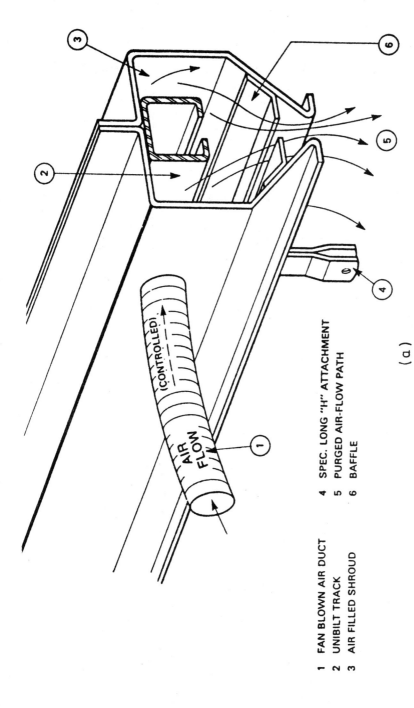

1 FAN BLOWN AIR DUCT   **4** SPEC. LONG "H" ATTACHMENT
2 UNIBILT TRACK   **5** PURGED AIR-FLOW PATH
3 AIR FILLED SHROUD   **6** BAFFLE

(a)

Figure 8.2 Two methods of protecting the conveyor; (a) air-flow protection scheme and (b) fully shrouded conveyor track. (*Courtesy of Jervis B. Webb Company*)

1 #20200 UNIBILT TRACK

2 FORMED ASTRAGAL SUPPORT STRIP

3 ASTRAGAL BACKUP RIVET STRIP

4 NEOPRENE (OR EQUAL) ASTRAGAL ⁵⁄₁₆" MINIMUM

5 SPECIAL LENGTH "H" ATTACHMENT

(b)

Figure 8.2 (Continued)

TABLE 8.1   Process-Time Chart

Process	Residence time, min		Max. fpm	Length, ft
Line load area	4	×	6	24
OT to washer	2	×	6	12
Power washer	5 2/3	×	6	34
OT to dry-off oven	2	×	6	12
Dry-off oven	5	×	6	30
OT to cool-down	12	×	6	72
Powder application	2 1/2	×	6	15
OT to cure oven	3 1/2	×	6	21
Cure oven	15	×	6	90
OT to cool-down	15	×	6	90
Unload area	4	×	6	24
OT to load area	6	×	6	36
				460

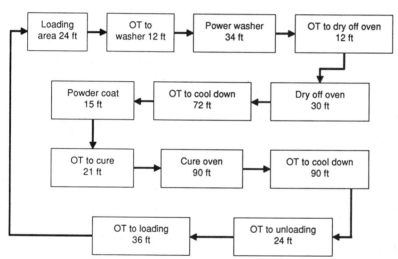

Figure 8.3   Block diagram of a conveyor layout.

present finishing area. Special items such as lubrication equipment, drives, and take-up units may require some small design accommodation, but your conveyor engineer will help you fit any of these items into your system at the proper location.

Laying the system out on graph paper

Now you have enough information to make a trial layout of the powder application equipment on graph paper. After establishing a scale, it's simple to make rectangular cutouts in block form and place them on graph paper to replicate various floor plans of your proposed finishing area. Be sure to include objects like pillars, floor drains, etc., in your plan. Move the cutouts around until they fit within the proper places on your proposed finishing line. Figure 8.4 shows what a graph-paper layout might look like with cutout simulated components in place.

There are common universal conveyor symbols used within the industry to identify various system components. These symbols are shown in Fig. 8.5. Use them in your design sketches.

Establishing the conveyor path

When you make your graph-paper layout, there are a few things you must consider when placing components of the conveyor path:

1. Parts and fixtures, especially long parts, need special consideration at corners. Chain-type material, articulated hangers, and moving load bars with trolleys on them can facilitate matters here. The

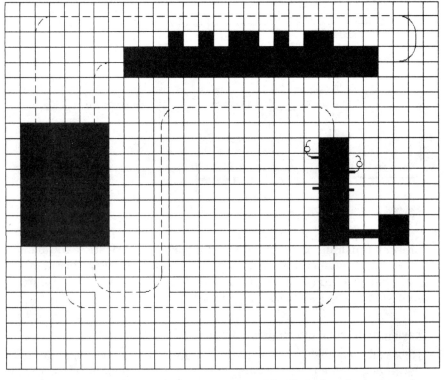

Figure 8.4 Graph-paper layout of a conveyor system. The block-shaped cutouts can be moved around on the graph paper until they are properly located, and then the conveyor path can be drawn.

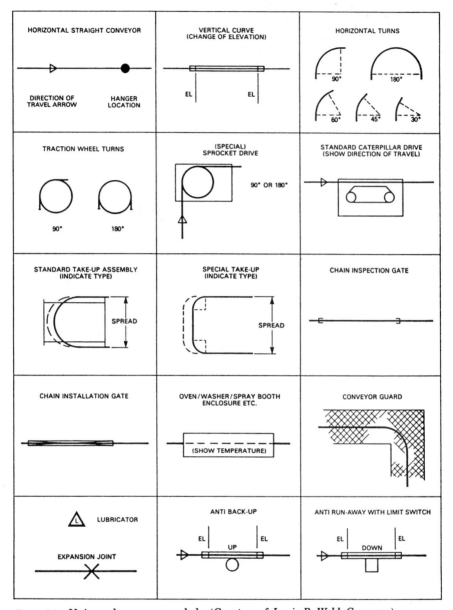

Figure 8.5 Universal conveyor symbols. (*Courtesy of Jervis B. Webb Company*)

important thing is that the radius of the horizontal turn should be large enough to accommodate the part. Figure 8.6 shows a method used to enable long parts to negotiate turns. Figure 8.7 shows some typical horizontal-turn arrangements, and Fig. 8.8 shows a way to handle vertical curves.

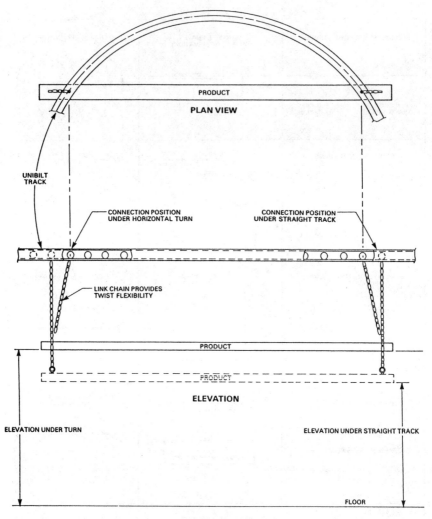

Figure 8.6 Making proper turns with long parts. The sketch is exaggerated to show true lifting reaction. (*Courtesy of Jervis B. Webb Company*)

2. After turning corners, extra long parts (long, that is, in the direction of conveyor travel) need plenty of distance to straighten out before entering equipment. Figure 8.9 shows long parts turning, both properly and improperly.

3. Frequently, a conveyor will be elevated as it carries the parts from one piece of process equipment to another. These elevated areas need safety caging to protect the people below if the parts should fall

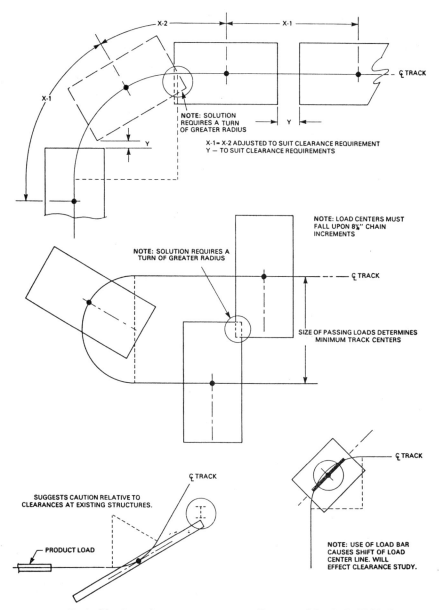

Figure 8.7 Typical horizontal-turn arrangements. (*Courtesy of Jervis B. Webb Company*)

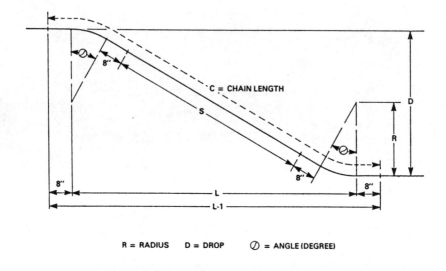

R = RADIUS   D = DROP $\oslash$ = ANGLE (DEGREE)

"L" DIMENSION
$$\text{"L"} = 2R \text{ SIN } \oslash + \frac{(D - 2R + 2R \text{ COS} \oslash)}{\text{TAN} \oslash}$$

"L-1" DIMENSION
$$\text{"L-1"} = \text{"L"} + 1'4''$$

"S" CONNECTING TRACK
$$\text{"S"} = \frac{(D - 2R + 2R \text{ COS} \oslash)}{\text{SIN} \oslash} \text{ MINUS } 1'4''$$

"C" DEVELOPED CHAIN LENGTH
$$\text{"C"} = \frac{\pi R \oslash}{90} + \frac{(D - 2R + 2R \text{ COS} \oslash)}{\text{SIN} \oslash} + 1'4''$$

Figure 8.8 Making proper vertical turns. (*Courtesy of Jervis B. Webb Company*)

from hanging fixtures. Figure 8.10 shows what the caging might look like and how it might be attached.

4. Try to eliminate crossovers, that is, areas where one conveyor line crosses another. It may require an elevated area with accompanying safety caging to accomplish this feat.

5. Ovens are hot, and so are the parts leaving them. Design your line so that the powder coating enclosure is not too close to the oven and so that hot parts leaving the oven don't radiate heat on parts ready to be powder-coated.

6. Conveyors need drive units. They also need take-up units, some-

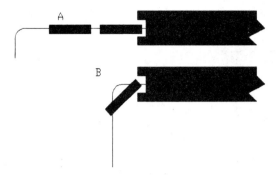

Figure 8.9 Making turns with long parts, both properly and improperly. In A, the parts have straightened out and will enter the oven straight on. In B, the parts are liable to touch the oven, which could possibly result in a reject.

times more than one, to allow for expansion and contraction within the chain as it changes temperature. Figure 8.11 shows a take-up unit. These units are composed of conveyor sections which open wider and stretch the chain as it expands. Take-up units can be controlled mechanically by pneumatics, weights, or springs or manually by a screw device.

Figuring the conveyor size

I have always said, "Don't send a boy to do a man's job, and don't use a 4-inch I-beam system to carry paper clips." To establish the proper conveyor size, first gather some pertinent figures:

1. Establish what your maximum loads will be at any one given time.
2. Establish the total weight on each hanging fixture and/or the total weight of part per linear foot.
3. Establish the total length of the system in feet.
4. Establish the total length of hanger in the direction of conveyor travel.
5. Establish the distances between hanging fixtures.

Then set up your equations. Let

W = maximum weight per part
P = parts per fixture
H = fixture weight
L = length of hanging fixture in direction of conveyor travel

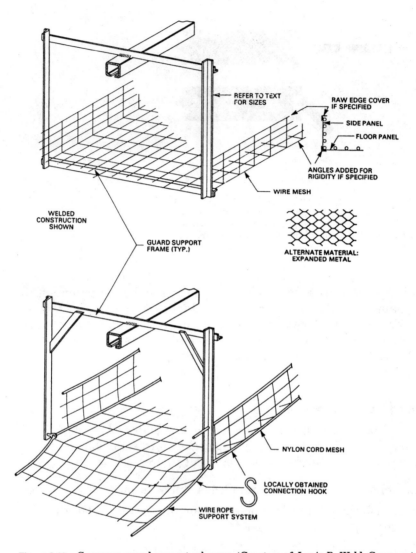

Figure 8.10 Conveyor guard support schemes. (*Courtesy of Jervis B. Webb Company*)

D = distance between fixtures
T = proposed system length

Figure 8.12 shows a simulated line and the calculation points W, H, L, and D. Using this as an example, where P = 3, we have

$$W\ (5\ \text{lb/part}) \times P\ (3) = 15\text{-lb load}$$

$$15\text{-lb load} + H\ (2\ \text{lb}) = 17\text{-lb load}$$

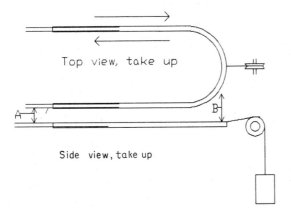

Figure 8.11 Conveyor take-up unit. The take-up section is the U-shaped (B) and is of slightly larger dimension than the actual conveyor track (A). The cable attached to the U section is also attached to a weight. As the chain expands or contracts, the U section slides out or in.

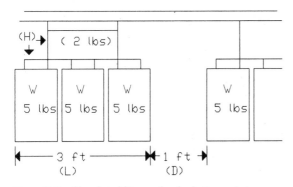

Figure 8.12 Simulated line and calculation points.

Length L of the hanging fixture is 3 feet (in the direction of conveyor travel). Distance D between hanging fixtures is 1 foot. So,

$$L + D = 4 \text{ ft}$$

From Table 8.1, we know that our system should be about 460 feet in total length. So,

460 ft/4 ft = 115 groups, hanging fixtures with spacing between

And

17-lb weight per load and fixture × 115 = 1955-lb total load

To this, add the weight of the conveyor chain. For a typical enclosed track chain, assume 2.9 pounds per foot:

$$2.9 \times 460 = 1334 \text{ lb}$$

The total load weight of 1955 pounds plus the chain weight of 1334 pounds gives us 3289 pounds. Spacing the 3289-pound load over the length of the conveyor system (460 feet) gives us

$$3289 \text{ lb}/460 \text{ ft} = 7.15 \text{ lb/ft load}$$

This number will be useful when a decision is made as to the type of conveyor to be used.

Now, can your ceiling joists handle the load, or will you need to use floor stands to hang the conveyor system? Going back to the 460-foot conveyor length traveling at 6 fpm, we can establish how many pounds of metal an hour will go through the oven. In the block drawing, the conveyor in the oven is 90 feet long. The parts will be in the oven for 15 minutes. So,

$$60 \text{ min per hour}/15 \text{ min in the oven} = 4$$

$$90 \text{ ft of conveyor in the oven} \times 4 = 360 \text{ ft}$$

$$L + D = 4 \text{ ft}$$

$$360 \text{ ft}/4 \text{ ft} = 90 \text{ packages}$$

$$90 \text{ packages} \times 17 \text{ lb per package} = 1530 \text{ lb}$$

$$360 \text{ ft} \times 2.9 \text{ lb (chain weight in oven per hour)} = 1044 \text{ lb}$$

Adding these last two weights, we arrive at a load of 2574 pounds of parts, fixtures, and chain per hour in the oven.

Chain-pull calculations

Manufacturers of conveyors know their business. After you've finished your sample layout, go over it with a competent sales engineer. Discuss the loads the system will be carrying. Most engineers can run a simple chain-pull calculation for you. When the final system is designed, they will run a proper one for you, probably on their own computer system. Among other things, this test will tell you if you need multiple drives and it can help specify the proper location for the drive units.

"Point-to-point" chain-pull calculations are frequently used when the conveyor system gets complicated. They are used to make certain the drive or drives are properly located. Elevation changes, both up and down, are taken into consideration, as are hanging fixture weights, coefficients of friction, vertical curves, and horizontal turns.

1. SHORT METHOD - GOOD CONDITION (OVERALL FRICTION FACTOR 2½ %)

CONVEYOR LENGTH —	= 355'-0"
CHAIN WEIGHT	= 2.9 LBS./FT.
<u>CARRIER @ 8.5 LBS. + ATTACHMENT @ 1.5 LBS.</u> 24" LOAD CTRS.	= 5 LBS./FT.
	8 LBS./FT.
<u>PRODUCT @ 44 LBS.</u> 24" LOAD CTRS.	= 22 LBS./FT.

TOTAL LIVE LOAD = 30 LBS./FT.

355 FT. x 30 LBS./FT. = 10,650 LBS. x .025% = 266 LBS. CHAIN PULL

LIFT LOAD (5 FT. + 5 FT.) = 10 FT. x PRODUCT @ 22 LBS. = 220 LBS.

486 LBS. TOTAL CHAIN PULL

2. POINT-TO-POINT CHAIN PULL CALCULATION — GOOD CONDITION

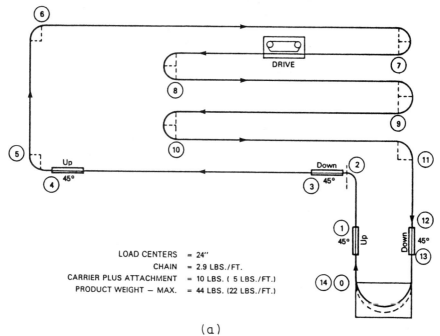

LOAD CENTERS	=	24"
CHAIN	=	2.9 LBS./FT.
CARRIER PLUS ATTACHMENT	=	10 LBS. (5 LBS./FT.)
PRODUCT WEIGHT — MAX.	=	44 LBS. (22 LBS./FT.)

(a)

Figure 8.13 (a) Comparison of methods of chain-pull calculation (for the same conveyor project) and (b) sample point-to-point chain calculation. (*Courtesy of Jervis B. Webb Company*)

Figure 8.13a and b shows a simple chain-pull calculation and a point-to-point chain calculation for the same hypothetical system.

Supporting the conveyor load

You know what your parts weigh, and it is easy to find out what your hangers and conveyor components weigh. But, chances are, the con-

SAMPLE CHAIN PULL PROBLEM
POINT-TO-POINT

$$\text{CHAIN AT 2.9 (LBS./FT.)} + \frac{\text{CARRIER AT 8.5 LBS.} + \text{ATTACH. AT 1.5 LBS.}}{24''} = 8 \text{ LBS./FT.}$$

$$\text{PRODUCT AT } \frac{44 \text{ LBS.}}{24''} = 22 \text{ LBS./FT.}$$

TOTAL LIVE LOAD   30 LBS./FT.

CHAIN PULL PER FT. OF LOADED CHAIN = 30x1½%* = 0.45 LBS./FT.

(*PREDETERMINED ROLLING FRICTION)

CHAIN PULL CALCULATION

```
0-1 — 17' x .45 =   8 + (5 x 22) = 118           x 1.025 = 121 LBS.
1-2 —  9' x .45 =   4 + 121 = 125                x 1.025 = 128 LBS.
2-3 — 12' x .45 =   5 + 128 = 133 – (5x8) = 93   x 1.025 =  95 LBS.
3-4 — 50' x .45 =  23 + 95 = 118 + (5x22) = 228  x 1.025 = 234 LBS.
4-5 —  5' x .45 =   2 + 234 = 236                x 1.025 = 242 LBS.
5-6 — 20' x .45 =   9 + 242 = 251                x 1.025 = 257 LBS.
6-7 — 54' x .45 =  24 + 257 = 281                x 1.045 = 294 LBS.
7-8 — 39' x .45 =  18 + 294 = 312                x 1.045 = 326 LBS.
8-9 — 39' x .45 =  18 + 326 = 344                x 1.045 = 359 LBS.
9-10 — 39' x .45 = 18 + 359 = 377                x 1.045 = 394 LBS.
10-11 — 36' x .45 = 16 + 394 = 410               x 1.025 = 420 LBS.
11-12 — 12' x .45 =  5 + 420 = 425               x 1.015 = 431 LBS.
12-13 — 11' x .45 =  5 + 431 = 436 – (5x8) = 396  x 1.025 = 406 LBS.
13-14 — 12' x .45 =  5 + 406 = 411               x 1.045 = 429 LBS.
```

355 FT. CONVEYOR LENGTH   TOTAL CHAIN PULL = 429 LBS.

EXPLANATION OF ALLOCATION OF FRICTION FACTORS

POINTS	SEGMENT OF SYSTEM	%*	POINTS	SEGMENT OF SYSTEM	%*
0-1	STR TRACK + 45° RISE	2½%	7-8	STR TRACK + 180° TURN	4½%
1-2	STR TRACK + 90° TURN	2½%	8-9	STR TRACK + 180° TURN	4½%
2-3	DECLINE AT 45°	2½%	9-10	STR TRACK + 180° TURN	4½%
3-4	STR. TRACK + 45° RISE	2½%	10-11	STR TRACK + 90° TURN	2½%
4-5	STR TRACK + 90° TURN	2½%	11-12	STR TRACK ONLY	1½%
5-6	STR TRACK + 90° TURN	2½%	12-13	DECLINE AT 45°	2½%
6-7	STR TRACK + 180° TURN	4½%	13-14	STR TRACK + 180°	4½%

(b)

Figure 8.13   (Continued)

tractor who built your building didn't know you were going to hang a conveyor from the joists. Because of that, it may be necessary to hang your conveyor from floor-mounted structures. Conveyor companies can help you with the design of these stands. Figure 8.14 shows a typical floor stand, which is normally bolted to the floor.

Figure 8.15 shows some methods of hanging a conveyor from the

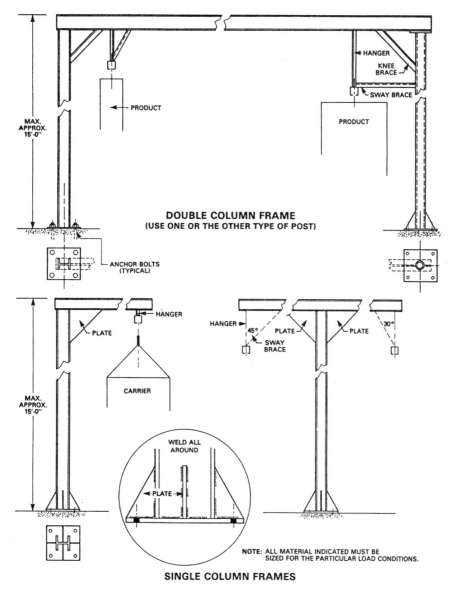

DOUBLE COLUMN FRAME
(USE ONE OR THE OTHER TYPE OF POST)

SINGLE COLUMN FRAMES

Figure 8.14 Typical conveyor floor-support schemes. (*Courtesy of Jervis B. Webb*)

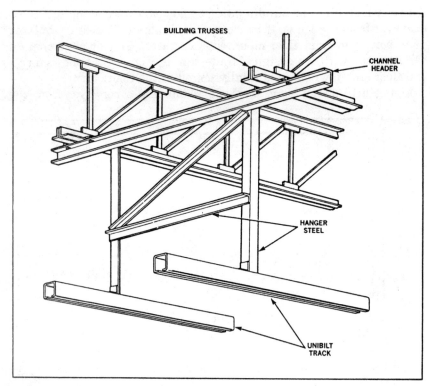

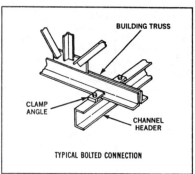

Figure 8.15   Methods of supporting a ceiling-mounted conveyor. Wherever possible, avoid welding superstructure and hanger steel to the building steel; use bolted type connections. On hangers and header steel, welded connections are considered more economical than bolted connections, but either type can be used. Conveyor loads and spacings govern the selection of hanger and sway-brace-angle sizes. (*Courtesy of Jervis B. Webb Company*)

ceiling. Professional installers know how best to hang your conveyor, and factory engineers can also help you.

Types of Conveyors

There are many types of conveyors available to handle your parts. Their load capacities are established by many factors. Total weight to be carried, elevation changes, factory conditions, total length of conveyor, quantities of curves, all have a bearing on the eventual size you will choose for your system. As you look at conveyor literature and talk to conveyor people, you'll begin to understand load limits and capacities.

There are two very general categories of conveyor systems—the enclosed monorail and the exposed rail. The enclosed monorail can be designed in the round tube or square tube style. The exposed rail system can be designed with I beams (which are available in 2-, 3-, 4-, and 6-inch sizes, or larger), flat bars, T rails, and round tubes.

Figure 8.16 illustrates some different conveyor styles.

Power-and-free systems

Power-and-free conveyor systems are more expensive when compared to straight conveyor systems, but they can save you floor space and labor costs. In fact, the uses of such a system are limited only by your imagination. Consider some hypothetical situations in which power-and-free systems could serve admirably:

- Your conveyor system will need to travel 25 fpm to provide the production rate you want from your system. About half of your parts can be coated automatically; the other half will need touch-up. It is almost impossible to touch up manually at 25 fpm; the simple solu-

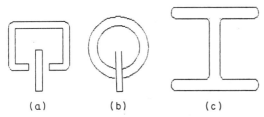

(a)   (b)   (c)

Figure 8.16 Profiles of different conveyor styles: (a) square-tube type; (b) round-tube type; and (c) I-beam type, which starts at 2 inches and goes to larger sizes for larger and heavier parts. The pendant of the square- and round-tube types is sometimes offset.

tion is to use a power/free system, where the B enclosure conveyor speed is approximately 12.5 fpm. Figure 8.17 gives you an idea how such a system, with proper spacing, might work.

- Oven curing space is at a premium. With a power-and-free system you can stack parts and increase your oven capacity. Figure 8.18 shows an example of this.

- You spray four colors; about 25 percent of your production is devoted to each color. You can expedite the system with the use of a power/free conveyor. Figure 8.19 shows an example of such a system.

- Finished parts go to several different assembly stations. Occasionally there is some confusion, and the wrong parts go to the wrong station. With a system such as that shown in Fig. 8.20 parts can be stored on seven lines. Sensors place the proper hangers on the proper storage lines; an operator can release a hanger from any storage line at will. Hangers and parts then go to the proper work station. This labor-saving device also prevents finished parts from becoming damaged while waiting to be assembled.

- Your parts must be assembled at a point some 1500 feet from where they were finished. Power/free systems can move parts at very high speeds to distant locations. Figure 8.21 illustrates a system where parts are manufactured and placed on a conveyor at a line speed of

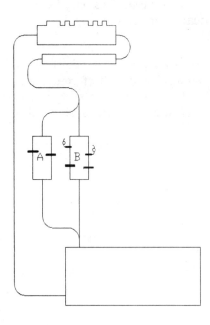

Figure 8.17 Power-free conveyor systems. Enclosure A is an automatic spray system (25 fpm); line B is an automatic/manual touch-up system (12.5 fpm).

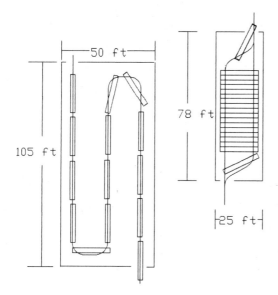

Figure 8.18 Stacking parts in a power/free system, as opposed to conventional conveyor design.

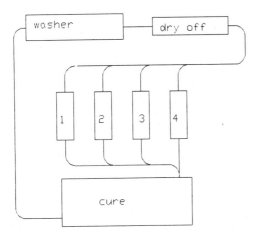

Figure 8.19 Use of power/free conveyor in a color coating system. Coating enclosures 1, 2, 3, and 4 each spray a separate coating. Fixtures are coded so that they go to the proper enclosure.

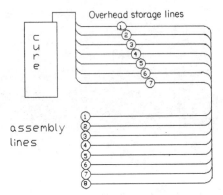

Figure 8.20 Storage rack on a power/free line. Parts leave the cure oven and go to seven storage lines, one for each of the seven parts. As a particular part is needed on one of the eight assembly lines, it is called for and sent to the proper line. No parts are manually handled or damaged.

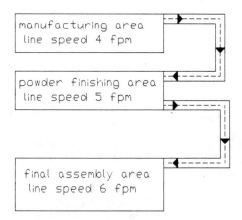

Figure 8.21 A variable-speed power/free line. Parts travel in a tunnel between buildings at 40 fpm.

4 fpm to be transported a distance of 1500 feet to the finishing building. Here they are finished on a powder-coating line traveling at 5 fpm. They are then sent to the final assembly and packaging warehouse where they travel at 6 fpm. In between buildings, the enclosed conveyor travels at 40 fpm. When parts arrive in any building, they are placed on storage lines, until needed.

Programmable Controllers

Programmable controllers can be used to control power/free systems and can be integrated with other controllers to handle other portions of your system. For example, they can turn your powder coating enclosures on and off as loads approach the entrance of the enclosure. When set up to do so, they can batch groups of parts through booths, rather than sending single fixtures through, thereby getting better powder distribution on your parts. They can control the timing of the hanging of parts within the oven, and they can control quantities of parts on particular lines. When set up properly, they can keep an accurate account of what is taking place within the system at any given time. They can start and stop an oven and constantly check its temperature; they can control the temperature, concentration, and pH level of a power washer; in fact, they can be used to monitor any component within the finishing area.

System Maintenance and Lubrication

An owners' maintenance manual or an instruction manual will be included with your system. These will give you some general information on maintenance. Like anything else, conveyor parts wear slowly. Your conveyor supplier will tell you what to look for and how to look for it.

High-temperature cure ovens require that your conveyor chain be supplied with a lubricant to withstand their temperatures. Get a good system to do this job right, or you'll be budgeting a lot of money each year for repairs. Your regular maintenance crew will have trouble keeping up with chain lubrication, so get an automatic lubrication system. There are companies that specialize in electro/mechanical/pneumatic systems which lubricate your chain at regular times. They actually count the times a given piece of chain passes a certain point, and then lubricate on schedule.

These units also inform you when they need maintenance. Figure 8.22 shows various aspects of a lubricating system.

Product Protection

Some manufactured products have extremely important cosmetic finishes. A blemish here or there could cause a reject. A sanitary shield can be used to eliminate any contamination to the finished product. As you can see in Fig. 8.23, the drip pan will collect drifting powder dust and lubricants that might fall from the conveyor.

Getting Maximum Use of a Conveyor System

Not knowing exactly what you will want to carry with your system, it's difficult for me to tell you how to plan for a finishing-line conveyor

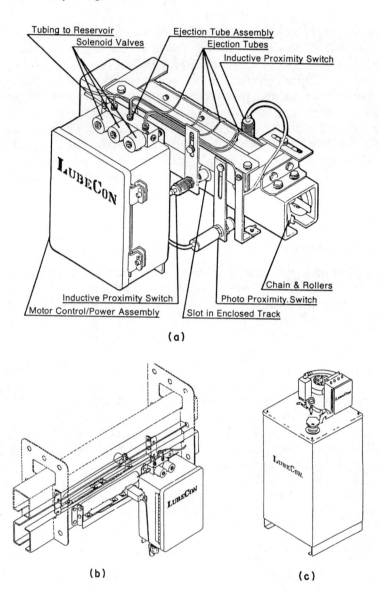

(a)

(b)   (c)

Figure 8.22 Automatic lubricating systems. (a) Model LCA-614/ET 3 Automatic Chain Lubricator; (b) Model LCA-614/FC 1 + 2 Free Trolley Wheel Lubricator; (c) centralized pumping station; and (d) Model LCA-614/ET3 with FC 1 + 2 (*Courtesy of Lubecon*)

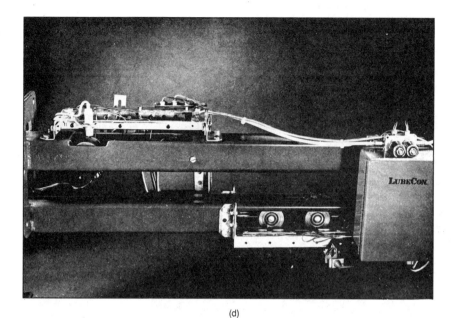

(d)

Figure 8.22 (*Continued*)

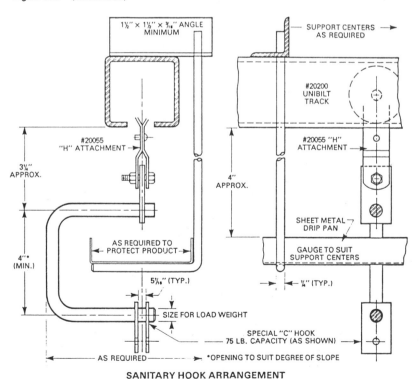

SANITARY HOOK ARRANGEMENT

Figure 8.23 A sanitary shield system. (*Courtesy of Jervis B. Webb Company*)

Figure 8.24   Optimum conveyor layout.

Figure 8.25   Conveyor storage racks. (*Courtesy of Jervis B. Webb Company*)

that's fully loaded and running all of the time, yielding maximum production. If not done properly, loading and unloading can cause conveyor downtime. Production can't be controlled if the line is constantly starting and stopping. If possible, finishing-line conveyors should be loaded at the point where parts are manufactured and unloaded at the point where they are finished, cured, and cooled. Invariably, fewer finished parts will be rejected because of damage if they are conveyed to the point of assembly or packaging. (See Fig. 8.24.)

Remember, the conveyor process is cheaper and more efficient than manual handling of the product. Consider some secondary uses of your conveyor system. Use a separate conveyor line to store your extra, unused hanging fixtures on. Think about it; it's a good idea to keep them off the floor. The unused area above the power washer or possibly the oven can make an excellent storage area. Figure 8.25 shows an example of this. Another use for a conveyor: Install a length near the loading area of the system; preload small parts off-line, then hang the loaded fixtures on the powered line.

9

Ancillary Equipment

Overview

Ancillary equipment includes the support equipment that comple-
ments the powder finishing system. I would like to talk about a few of
these pieces.

Stripping Equipment

We would all agree, at least those of us who have tried, that it is difficult to remove cured powder. There are many ways to strip; the question has to do with exactly what it is you want to strip.

If you simply want to remove cured powder from steel, it's easy. If
you need to strip zinc castings, aluminum castings, or aluminum extrusions, there may be a problem because of the heat required in the
stripping system.

Another point to consider is cost. There is not only the original cost
of equipment, there is the cost of the heat that is required in most
stripping methods. Along with heat, some methods use a media material or a chemical. Will your only cost there be the media or chemical,
or will there be other charges associated with getting rid of the sludge
or of the chemical when it's lost its potency?

How much stripping will you be doing? Will yours be a once-a-week-
type operation, or will you be stripping a few hours each day?

Based on my experience, I have a favorite method or two. For instance, I have used gas burn-off ovens with an internal water supply
to remove cured powder from parts and hanging fixtures with great
success. Opponents of burn-off ovens talk about the residual ash on
hanging fixtures and parts when they leave the oven. I agree; there is
an ash. But I have seen it removed very successfully by quenching
parts and fixtures in plain water as soon as they leave the oven.

Figure 9.1 shows a typical burn-off oven. The parts and/or fixtures are soaked in high temperatures (600–900°F). The time cycle can range from 30 minutes to a few hours. The more frequently fixtures are stripped, the more quickly they will strip in the oven. The process is simple. All of the effluents in the process are heated in a secondary burner chamber of the oven, which takes the products up to some 1600°F or more. These products then turn to carbon dioxide and steam, which is exhausted from the building.

An alternative to burn-off is chemical stripping, which takes a stripping tank and a rinse tank. The stripping times depend on strength of the chemical, the powder buildup, the generic type of the powder, and the heat. Though there are costs and problems—what to do with the

Figure 9.1 A burn-off oven, the Bayco BB-78. (*Courtesy of Bayco*)

sludge residue, whether the rinse water can be drained without treat-ment—there are times when chemical stripping is almost mandatory.

A plastic media, abrasive blasting, or abrasive stripping methods can be used. I have no personal experience in this area, but I under-stand that it works well. The plastic media is made of a thermoset plastic material and is available in different hardness levels and par-ticle sizes. It is possible to recycle the material several times during its use. Proponents of the process state they can strip a coating one layer at a time. Figure 9.2 shows the plastic media size as compared to a pencil point; Fig. 9.3 shows the cabinet-type machine used for the process.

A few other methods are also available, and all of them work well. If you are interested, investigate. The original equipment cost must be considered. So too must the operating costs, along with your volume and the nature of the substrate material. Just be sure to test the pro-cess and run a total cost analysis before you purchase any equipment. If the stripping job is minimal, you may want to have it done in a shop that specializes in stripping.

Air Compressors and Driers

Sometimes—for instance, on very small systems—dedicated air com-pressors are not necessary. Sometimes—for instance, on very small systems—refrigerated air driers are not necessary. I have seen many successful systems run without either. It would be difficult to prove how much of an efficiency increase there would be with a dedicated air compressor and an air drier on a given system. I will say this: I do feel clean, dry, cool powder flows better through equipment than does

Figure 9.2 Plastic blasting media.

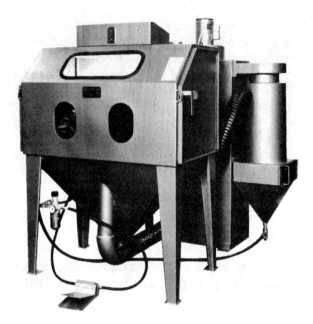

Figure 9.3 Machine for stripping with plastic media.
(*Courtesy of Aero Blast Products*)

warm, damp, oily, lumpy powder. As powder travels through equipment it creates friction. Friction creates heat. Heat is used to cure powder. Refrigerated, clean, cool, dry powder will invariably keep your internal parts, pumps, hoses, guns, nozzles, and controls operating better and cleaner.

You cannot believe the contamination on the inside of a length of pipe carrying air. Add to this the many forms of contamination created when the pipe is cut and threaded. I can recall a number of times when insignificant micron-sized particles of debris gained entrance to devices attached to powder coating application systems, causing production-line downtime. In one instance, a simple control had to be flown to the job site before production could be resumed. *You* may want to take such chances, but I wouldn't bet *my* job on it. So let's just say that a finishing system should have its own dedicated air compressor and drier. In most plants, compressed air is supplied from one or more compressors located in a special "compressor room," which can be next to the finishing area or a couple thousand feet away. Most powder application people can tell you exactly how much air you'll need. They can supply the compressed-air equipment and a matched refrigerated air drier system to go with it. Let them do it. They know what you need.

Along with most powder equipment, especially at the control console, you will sometimes find a cartridge-type filter. Do not believe for a minute this will take the place of a refrigerated air drier, filter system, and coalescer. But equipment companies know what happens when new systems start up and foreign matter within compressed air pipes travels toward air-controlling equipment; the cartridge filter simply represents an insurance policy, a last line of defense.

Other than the air needed for the application equipment, some small quantity of compressed air is frequently needed for operating equipment on the conveyor system, particularly a power/free system. Where there is UV flame-detection equipment in automatic powder coating enclosures, a small jet of air is generally used to keep the "eye" of the system clean.

A Powder Sieve

To sieve or not to sieve? If you discuss powder sieving with ten different people, you'll probably get ten identical answers. They'll all say something to the effect, "There are times when you should sieve and times when it won't be necessary." Ask your powder supplier if you will need a sieve in your system. Get a definite answer, one way or another.

Many systems operate without sieves. But there are those times when sieving is essential. Normally powder is in good shape when it comes to you, but if your warehouse people take the palletized boxes and store them in a warm place for a while, the powder could get lumpy. In traveling through the feed hopper, pumps, hoses, and guns, this lumpy powder could conceivably clog the equipment.

Sieving works well when you are using clear coating materials. It also works well in combining reclaimed material with virgin material. It helps balance the particles, leading to better transfer efficiency.

In buying a sieve, remember that if it is to be involved in a color change, it will need to be cleaned. Find out exactly how long the cleaning will take. An extra half hour of cleaning with each color change may cause you to decide on multiple sieves or to turn to another solution.

Sieves can be easily integrated into a system. Different styles are made by different companies. Ask your powder supplier to recommend the best style for your system.

Clean-up Equipment

It's amazing how much dirt can accumulate in the finishing area, if you let it. If it's to operate well, your system will need constant cleaning and maintenance.

Shortly after your conveyor starts to operate, the automatic oiler may need an adjustment. The easy way to tell is to look on the ground directly under the chain at the low points of your inclines and declines. If there's oil there, your oiler needs an adjustment. If you don't make the adjustment there, the leaking oil will progress to the bottom of the enclosed track, where it will hang in droplets. These must be wiped off occasionally before they affect your finish. They won't hurt the parts to be pretreated because the washer chemicals should remove them. It's the area after the part leaves the washer and before it's cured that you must be concerned with because the simple truth is that powder and oil do not mix very well. The important thing is, catch the oil early and keep it off the parts and the plant floor.

You will find that overspray powder tends to collect on the walls of your coating enclosure. Part of this is due to the electrostatic charge given the powder particles as they leave the powder gun, and part of it is from the force of the air propelling the powder when it leaves the gun. I have seen an interesting item on the market which "deionizes" the clinging particles as it sprays them with compressed air. The idea works fairly well. If it has no electrostatic charge, the overspray powder will be pulled from the enclosure to the reclaim or collection system where it can be recycled.

No matter, the interior of your powder coating enclosure will inevitably collect some overspray—how much depends upon the original design of the equipment. It's easy to squeegee this powder toward the reclaim system. If you plan to change colors, it's a good idea to use the squeegee every 30 minutes or so. Remember that reusing powder left clinging to your enclosure walls can reduce the number of boxes of powder you'll have to add to your feed hopper.

Squeegees are also handy tools to have around a power washer. They help keep the floor clean and dry, despite liquid spills.

Dust will be a constant in your system, but whether it's powder dust or just dust in the air, it's a problem you can deal with. Let's look at the real world: Your powder coating enclosure was designed to keep overspray powder within the enclosure area and to send most of it to the collection or reclaim system. Despite this, the new loading dock door you installed—you know, the one where the prevailing wind pours in—tends to unbalance the coating enclosure. Until someone readjusts the collection-system dampers, powder will escape into the powder application room or into the area around the powder coating enclosure. Simple maintenance will control the problem.

Prior to installing the powder coating enclosure on a concrete floor, clean the floor thoroughly and apply a good floor sealer. Put a couple of coats of sealer on the floor. The reason for the sealer is simple: to counteract the porosity of the concrete and to keep the tiny particles of powder and other dust from being trapped in a porous floor (or wall).

Let's say yesterday you used red powder. Today you're using white powder. Let's say the guy you hired to move pallets around walks by the coating enclosure in which you're now using white powder. His spongy soles expel air as his feet move across the floor. All of a sudden your white parts have red measles; you have contamination and next you'll have rejected parts. Don't forget to seal the floors; it's important.

I like to see a good mop, along with a wringer-type mop bucket, assigned to the powder application room only. It only takes a few minutes a couple of times a day to damp-mop the area. It eliminates much of the dust, keeps contamination down, and gives the feeling of working in a nice atmosphere.

For color changes and for heavy-duty cleanup, a good canister-type vacuum cleaner is a must for any powder application area.

Each manufacturer will tell you how its particular equipment should be cleaned. But the above-mentioned items, along with a couple of blowdown air hoses for cleaning powder guns, pumps, and hoses are the basic tools for keeping your plant operating efficiently.

Testing Equipment

There is equipment to test almost anything in the finishing industry. Equipment for testing cured powder films ranges from a simple mil-thickness tester to a very sophisticated record-keeping machine that gives out readings on a tape and also stores the readings internally.

Periodically during the day, finishing-line personnel should do their own testing. Portable units about the size of a cigarette package do an excellent job of monitoring production processes, but your quality-control people may need more sophisticated equipment in a laboratory. I know it's possible to buy equipment for in-house testing of any normal industry procedure.

You will need some testing equipment for your power washer. Everyday titration equipment is usually furnished at no charge by chemical suppliers, along with record-keeping charts. If a close pH check is needed, there are both simple, handheld, inexpensive instruments and very sophisticated instruments. There are any number of systems furnished by chemical suppliers to automatically test the chemical in your tanks and add more chemicals as needed. The idea is to reduce the time your employees spend in titrating your system.

As mentioned in Chap. 7, there are instruments that can be attached to your hanging fixtures to record oven temperatures at several locations on the hanging fixture during the trip through the oven. The instruments will display the temperatures on a recording tape, giving you the complete heat record for every foot of travel in an oven.

If I ever buy another piece of equipment, it will be one of these units. No matter how many different weighted loads go through your oven and no matter how many different brands of powder you use for different jobs, you can always speak with certainty about your oven temperatures and time cycles, if you can refer to recent oven testing charts.

Powder manufacturers have almost everything, and if you feel you need some test equipment and are not certain what to buy, talk to your supplier.

10

Mil Thickness and Other Tradeoffs

Overview

People new to the powder finishing industry hear about many exciting things. Eventually, these people will form their own opinions about each new development. But let me lay some groundwork on the issues here and offer an expert's opinion.

Powder Thickness Tradeoffs

The first comments the novice in the powder industry hears generally come from a powder manufacturer's representative. The representative will talk honestly and sincerely about a sample of cured powder coating on a flat panel. The representative will have measured the mil thickness in several areas and noted the mil-thickness average on the panel. Let's say that 1.2 mils is the figure given. The novice and the representative might then engage in a discussion that goes something like this:

NOVICE: What kind of coverage can I get with this powder?

REP: You'll get about 117 square feet per pound at 1 mil.

NOVICE: How thick should this material be sprayed?

REP: You'll want to spray it at about 1.2 to 2.5 mils.

NOVICE: What impact resistance can I expect?

REP: Expect about 100 inch-pounds.

NOVICE: How about salt-spray resistance?

REP: It'll give you about a thousand hours.

NOVICE: How much does this powder cost per pound?

REP: $3.45 a pound on blanket orders of 10,000 pounds, with 2000-pound releases over the next 3 months.

The discussion ends, and the novice is now a distinct danger to himself and others with all his new-found knowledge. The next day a powder-application-equipment representative comes to talk to the novice. In demonstrating the equipment, the representative sprays some powder on some blank panels. The novice notices for the first time the wraparound effect of electrostatics and queries the representative, who explains the phenomenon to him, then sprays some test panels with a hand gun so the novice can show management what can be done to flat panels with powder. The cured samples come out with an average film thickness of 1.4 mils. Because of the wraparound electrostatic effect, the edge on the panel has a slightly heavier mil build. (Figure 10.1 shows such a hypothetical coated panel.)

The novice now thinks it through: "The powder man gave me a 1.2-mil average, and the equipment man talked about 1.4 mils. I think I'll put them together, divide by 2, and go with the average of 1.3 mils." He writes a report for management on what he's learned. In order to show management how thorough he is, he makes the report as detailed as possible:

> I have made some preliminary calculations based upon my meeting with the powder and the powder gun representatives. This electrostatic thing is a real miracle. You spray on the front side only, and the powder wraps itself around to the rear of the panel.
>
> The steel boxes we make have 5 square feet of surface on the exterior

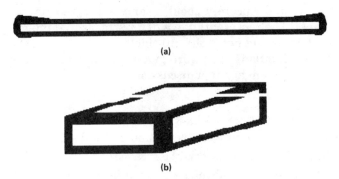

(a)

(b)

Figure 10.1   Coated flat panel. (a) The edge view of the flat panel shows approximately how the wraparound effect gives a slightly heavier mil thickness of powder at the edges. (b) A similar effect can be seen at the outside corners of the flat panel; the filled-in edges of the box represent the areas that will get a slightly heavier coating because of the electrostatic effect.

and the same on the interior. You'll see in the attached figure the box to be coated; note its corners and recessed areas.

The powder costs $3.45 a pound and I figure we can get by with a thickness of 1.3 mils. The powder salesperson says we will get 117 square feet of coverage per pound, so I'm figuring a material cost of $0.39 per metal

box.

$$\$3.45/117 \text{ sq ft} = \$0.02948 \ (\$0.03) \text{ per sq ft @ 1 mil}$$

The 1.3-mils coverage planned on the 10-square-foot box works out to

$$10 \text{ sq ft} \times \$0.03/\text{sq ft} \times 1.3 \text{ mil thickness} = \$0.39 \text{ cost per box}$$

At 50 boxes a day × $.039, we should have $19.50 in total material cost a day. I think we had better buy this new powder system now and get rid of our solvent-based paint.

Some very simple mistakes have been made in all sincerity by our novice. For example,

1. The powder usage he projects per day would amount to

$$10 \text{ sq ft} \times 50 \times 1.3 = 650 \text{ sq ft}$$

At 117 square feet per pound, this works out to 5.6 pounds per day usage. The price quoted per pound was based on the use of 10,000 pounds in 3 months. It would take our novice about 1786 workdays to use the 10,000 pounds of powder.

2. Despite his assumption, the electrostatic effect will not give 1.3 mils across the entire surface.

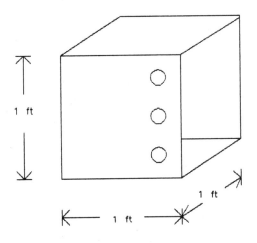

3. He has not allowed for overspray, which will be a factor whether the products are coated manually or coated automatically followed by a manual touch-up.

As a result of all of the above, the entire price structure is not exactly the way our novice understands it. No one has misled or lied to him. He just moved too quickly. It takes much more information to make the decision than our novice had when he filed his report with the boss.

He should have considered the tradeoffs, for instance. First of all, the powder representative probably got the test panels from a laboratory where they had been coated under laboratory conditions to show to their best advantage to the potential customer. In no way would the laboratory conditions match production-line conditions in the novice's plant.

Then too, if the quantities to be coated were large, for example, 500 boxes an hour, there would most certainly be some automatic equipment involved. The Faraday cage effect would tend to increase the mil thickness on the edges and outside corners. The insides of the box would also present a different mil thickness; the corners may be a little lighter, whereas the interior walls a little heavier. In any event, keeping the mil thickness at a good level would take planning and adjustment by the equipment operators.

Had our novice taken the time to get the powder and the equipment representatives together in a powder equipment laboratory and run some tests prior to writing his report, he might still have a job.

Granted, I have presented a hypothetical case here. But very recently I talked to a "novice" who was prepared to forge ahead on a project, thinking he could get a 1.5-mil thickness on a complex product that was to be finished on a high-speed production line. No one had lied or misled him; he simply misinterpreted the facts given. You cannot, and should not expect to, do x pieces an hour, all at exactly the same mil thickness, whether you use automatic or manual equipment. The automatic equipment will probably be the most consistent once it is up and operating, but be prepared for the electrostatic effect that will create mil-thickness differences on edges and in corners. Remember that powder may spray very well on flat sheets placed closely together on a finishing line, but in reality part sizes will vary as production changes.

Even given large, flat surfaces, your automatic guns, while faithfully spraying the desired mil thickness in the center of the part, will give a slightly heavier coat at the edges and in the open spaces between parts. Figure 10.1 showed an enlarged view of an edge of flat sheet; there's nothing wrong with the spraying equipment here; it's just the nature of the electrostatic effect.

And we haven't even discussed what happens when you're spraying not just flat panels, but complex parts of one kind or another.

There are many tradeoffs in production finishing, and the slightest of changes, which may not affect the quality of the end product, will no doubt affect the cost of the end product.

Tradeoffs Based on the Substrate

Another tradeoff to take into consideration is that involved in the metal used in testing. Suppose the powder salesperson visits another novice and displays several panels, each with a large X scribed across its front. The representative explains that these panels are coated with the same powder the novice is going to use. These panels have been subjected to 1000 hours in a machine which alternately sprays them with salt solutions and then subjects them to UV rays, a very legitimate test that is used and understood by most people in the finishing industry. The test shows, among other things, the condition of the coating and the amount of creepage under the cured film by the rust formed on the scribed surface. It's an excellent test and a benchmark most of us work by.

What the novice may not realize is that the panels themselves have been especially prepared with an aqueous pretreatment in a five-stage (or perhaps a seven-stage) power-washer system. Nor does the novice know that the metal used in the panel is a 10/10 steel, and therefore is completely unaware of the differences to be allowed for when the process is introduced to the novice's own plant, which uses a three-stage power washer on secondary steel bought from all over the world.

Powder manufacturers are ethical; when they discuss test results, they are talking about the same tests other powder manufacturers use, based on a common benchmark. The point is, our novice will have to take all the differentials introduced by the process in the novice's plant into consideration.

System-Conversion Tradeoffs

In the consulting field we see many problems that arise from converting to powder. They're not bad problems; they're problems that can be easily avoided by careful planning.

Say that a company manufactures a product that for many years was finished with lacquer. A mil thickness of "about 1 mil" was standard. After the product was finished it was subjected to final assembly, which included the actual postforming on a high-speed production line.

Then along came powder coating, which was to be applied at about 1.5 mils, per manufacturer's specifications. But the people within the company who had made the decision to convert to powder coating insisted on a pencil hardness of 4H. To achieve this, a heavier mil thickness was required, though that thickness had never been required before. When the product was now subjected to postforming, two separate problems surfaced. First, when the part was bent to form, the coating tended to crack. And occasionally, whether because the hanging fixtures were not filled or because hooks were broken from the fixtures or for any other reason, the part received more powder and therefore a heavier coat than was normal. The dies used in postforming were so tight, the parts would not now fit. This led to the rejection of many parts.

Hypothetical problems, maybe, but they illustrate the tradeoffs that have to be weighed in the testing program prior to converting to powder. Working with any new material will give new results. Sometimes the new results create new problems, which must in turn be confronted and solved. If the committee approach advised in Chap. 2 is used in investigating a powder coating system for your company, it should give you input from all departments as to the impact of a new system. And input at the beginning should eliminate any surprises during the start-up process.

Hanging Fixtures

Overview

If a statue were to be erected somewhere for everyone who designed a hanging fixture actually used in the powder coating industry, the building would come tumbling down from the excessive weight. If, however, there were statues dedicated only to the people who had designed *good* hanging fixtures, there would be no problem with numbers or weight.

One of the first hanging fixtures in North America, historically speaking, was the rope looped over a tree limb. The purpose of this fixture was to dispatch the "bad guys" from this world. Believe it or not, some people still think this is all there is to making hanging fixtures. One hanging fixture (tree-limb type), one body. One hanging fixture (finishing-system type), one part. Speaking of so-called hanging-fixture designers, some of them should be hung from a tree limb for the poorly designed fixtures they have released into the market. Some of these fixtures, when filled with parts, collect more powder than the parts to be coated.

Some of the people who have in the past made fixtures for wet-paint systems still use the techniques they used in those fixtures. They think that for good coverage the fixture or rack should turn, to make certain all parts are coated. They don't realize that electrostatic principles, not the "garden-hose principle," are at work here and that powder will "float" onto parts. Most powder coating shops use the "straight-line" technique, with no turning of the fixture in the powder coating enclosure. So, if you're going to talk to a professional, talk to people who make hanging fixtures for powder systems.

Throughout this chapter, we'll talk about hanging fixtures with one or two vertical rods between the fixture and the conveyor itself. A good solid connection is required at these points to prevent excess swinging and a poor ground. The connection at these points is so important that it will be mentioned throughout the chapter.

Designing the Fixture

If you have the time, plan your own fixtures. If you don't, talk to the real pros, the ones who make powder fixtures for a living. Or talk to people who powder-coat for a living. Job-shop people depend upon their skills in racking and fixture design to survive. With all their experience, they can often visualize in their minds exactly how they will fixture a part they are quoting to finish. Sometimes, out of necessity, they come up with a new hanging fixture, or perhaps a new type of hook attachment for the universal hanging fixture. I have been in job shops where there was no need for concern because the quantities were small, the parts were big, and the price was right. However, when quoting against competition for a large job that involves many small parts, these pros know that the profit often lies in the hanging of the parts.

What ingredients go into the design of a good hanging fixture?

1. *Simplicity:* A complex fixture will probably cost much more money to make than a simple one. The object is to coat the parts, not the fixture. So keep it simple.

2. *Accessibility:* Automatic spray guns are blind; they don't have the ability to "see" exactly how you hang parts. "Shadows," as they are called, can hide a portion of a part from the automatic guns. And shadows cause rejects. Presenting the parts uniformly and properly will give you consistency in the finish.

3. *Placement:* Again, because automatic guns cannot "see" your parts, place them as intelligently as possible in front of the guns. The more parts in front of the guns, the better the distribution of the sprayed powder, the less the overspray, and the better the transfer efficiency.

4. *Consistency:* If you make fifty fixtures to hang certain parts, then all fifty should hang parts the same way and at the same elevation. When all parts are hung at the same level, at the same position in front of the automatic spray gun, then equal transfer efficiency can be achieved.

5. *Universality:* If it is at all possible, use one basic or universal fixture, with multiple adapters for different parts. Some simple examples of a universal hanger with adaptable hooks are shown in Fig. 11.1.

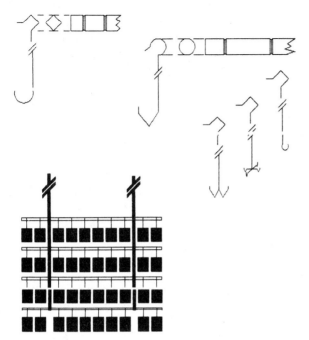

Figure 11.1 Two types of universal fixture, one using a round bar and one using a square bar. These types as well as others are available on the market; if you are hanging a variety of different parts, you may need several different types of hooks.

Units similar to this are available on the market. Fixture manufacturers will work with you on exact designs to meet your needs.

6. *Solid connections:* If it is at all possible, completely hide the point of contact between the fixture and the part. Protecting this point from collecting powder will keep a good ground indefinitely. One of the most perfect fixtures I have ever seen was one welded with small bolts that threaded into the part being hung. The bolt thus held the fixture while also acting as a mask for the threaded hole in the part. A perfect ground and a perfect mask, all on one fixture.

7. *Light weight:* Design the fixture so it is light in weight and yet strong enough to withstand mishandling in loading and unloading. Remember, extra weight in your oven wastes fuel and robs heat from your coated parts, which need the heat to cure.

8. *Small openings:* The smaller the open space on the fixture, the less powder the fixture will collect. A clean, grounded fixture, like your clean, grounded part, acts as an antenna for the powder to collect on. A clean, well-grounded fixture that collects powder is robbing your

parts of that powder; the result is a shadow or even a bare spot. Powder collecting on your hanging fixture is a waste of money. It's a given fact that powder will always collect on a fixture, but try to minimize the amount of that powder and save your money.

9. *Strength:*   Eventually you'll probably want to strip your fixtures, or at least part of them. So build them strong enough that they can handle any excessive or abusive handling they'll receive while they're being stripped.

10. *Rigidity:*   A rigid hanging part will not swing in your washer. Nor will it swing while being coated with powder during its trip through the powder coating enclosure. A small, light part could swing under the spraying pressures of chemicals or powder, and too much swinging can dislodge parts from their fixture, causing a reject.

Special Considerations

It would be impossible for me to describe the perfect hanging fixture for you and your company here. Instead, I'll show you some of the ideas I have seen at work and let you develop your own fixture.

Rigid mounts

To begin at the beginning: How are you going to hang the fixture on the conveyor line itself? If you use a single point of contact, your fixture could swing unintentionally, causing parts and fixtures to get caught crosswise in the entrances or exits of washers, ovens, and coating enclosures. I have even seen a sideways swinging motion become so strong that it caused parts to fall off the line, ruining finishing equipment in the process. But even if the sideways swinging motion does not cause parts to fall off the fixtures, it can still eventually add up to a better than 60 percent reject rate because it interferes with an even, consistent spraying.

Figure 11.2 shows the swinging which will occur with a one-point attachment. A fixture hanging from a single point can twist in any one of the four directions unless it is solidly suspended at that single point. Long, lightweight hanging parts tend to swing when hung from one point. But a fixture suspended from two points will stay somewhat rigid and not swing as the fixture labeled in Fig. 11.2 would.

For this reason, I personally prefer a rigid fixture, which allows for no sideways swing. If done properly, a rigid design will not create a problem when vertical conveyor turns (rises and drops) are used.

top view of conveyor system

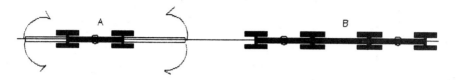

side view of conveyor system

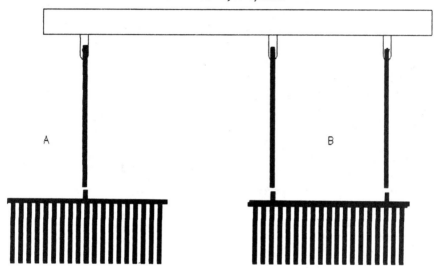

Figure 11.2 One-point attachment (A) permits the hanger to swing in the directions shown; two-point attachment (B) prevents the hanger from swinging.

On conveyor inclines and declines, loose hanging parts will continue to hang vertically, as will a singularly hung fixture such as the one labeled A in Fig. 11.2. On the other hand, the fixture labeled B in that figure will follow the angle of the conveyor incline or decline, be it 10° or whatever.

Close placement

Production systems dictate that parts be hung very closely. There is an important efficiency gained from placing hanging fixtures as close as possible, particularly when the parts are going through the powder coating enclosure. But you gain more than greater transfer efficiency of powder from the gun to the parts; you also gain closer control of mil thickness. Figure 11.3 shows how this works.

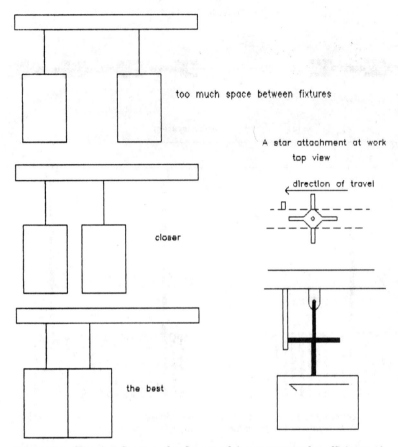

too much space between fixtures

A star attachment at work
top view

direction of travel

closer

the best

Figure 11.3 Hanging fixtures closely spaced increase transfer efficiency. A star device can rotate the part 90° when it is not within the coating enclosure. The star is shown in operation here; as the part approaches the vertical post, the post causes the star, and thus the part, to rotate.

Placing the fixtures extremely close together may be difficult, but it's made possible if the conveyor and the fixtures are set up properly. A device called a "star" can be mounted between the conveyor and the fixture, enabling the fixtures to be automatically turned 90° when traveling on inclines or declines. A star device can work alone, or with conveyor part-load bars; one is shown in Fig. 11.3.

Figure 11.4 shows you exactly what can be done to a fixture to improve its efficiency. We start out with a 72-piece fixture and then increase its efficiency to 107 parts.

It's also possible to hang parts on edge, as shown in Fig. 11.5. As-

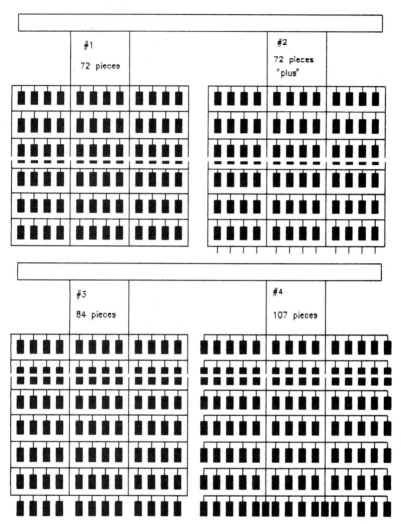

Figure 11.4   Increasing a 72-piece fixture to a 107-piece fixture.

sume in this case that the parts to be coated are flat sheets of metal, 10 inches by 12 inches, with about an inch between parts.

Hanging fixtures for extrusions and tubes

If your extrusions are to be hung horizontally, they can be as much as 20 feet long and still create no problem, provided your equipment is able to process these parts. Since power-washer nozzles can't "see"

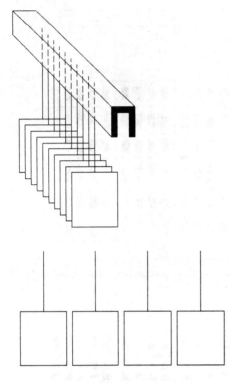

Figure 11.5 Hanging flat sheets on edge. Within limits, it is possible to hang parts as shown in the upper drawing, giving a denser loading, and hence a higher efficiency, than possible with the arrangement shown in the lower drawing.

your parts any better than automatic powder guns can, it'll be necessary for you to adjust the parts so the equipment can completely reach all surfaces of the part to be coated. The arrows in the extrusion profile of Fig. 11.6 indicate all exterior surfaces to be coated. This means the extrusions must also be pretreated in these areas.

The hanging fixture should be designed to alternate the position of the extrusions on each side of the fixture. Then there'll be plenty of room for the nozzles on the left side of the process equipment to get not only the front, the top, and the bottom of the extrusion, but also to apply some spray to the rear of the extrusions on the right side. Figure 11.6 shows approximately how this would work. The drawing shows three ways to hang extrusions horizontally. Drawing A of the figure shows a balanced fixture. As shown, a power washer would get some pretreatment to the rear of each extrusion, but probably not as much as it does to the front. If powder coating the rear was necessary, it would be very difficult, even with electrostatic wraparound. Drawings B and C show two possible ways of both pretreating and powder coating the exterior of the extrusion.

Hanging aluminum extrusions or tubing takes some planning. If they are to be hung vertically, they must be hung properly, that is,

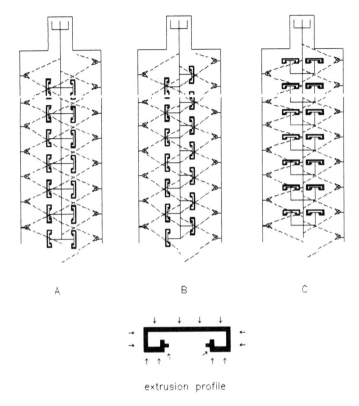

extrusion profile

Figure 11.6   Hanging extrusions.

hung so that they will stay rigid while traveling through the process and not hit each other. I have seen horizontal fixtures with long V-shaped pieces of spring wire fastened rigidly to the hanging fixture, with the wire's other end forced into the profile end of an extrusion, causing friction between the wire and the extrusion. Figure 11.7 shows how this is done with pieces of tubing, and approximately how the tubing would hang.

Grounding conveyor, hanging rack, and part

We can discuss proper grounding all day, and you will still not take it seriously if, as long as some powder clings to your parts when they are sprayed, you assume you are grounding properly. Not so. As you apply powder to your hanging fixtures and they go through the curing oven, you create resistance between the metal hanging fixture and your newly pretreated part. This resistance continues to build until eventually, when your operators turn up the powder feed, great clouds of

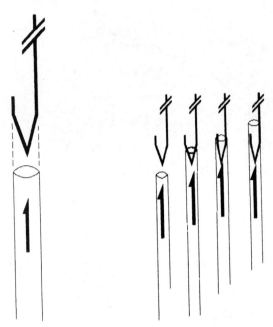

Figure 11.7   Tube hanging; A spring-loaded fixture is forced into the end of the tube; friction holds it in place. The same fixture can be used with aluminum extrusions.

powder billow through your coating enclosure. Then, when you call the gun manufacturer to complain, he'll be patient and sympathetic, and the first thing he'll ask you to do is to check the ground carefully.

Try to remember, a chain is as strong as its weakest link. The weakest link in the chain between a good ground and your part is the point of contact between your hanging fixture and your part. Usually the actual point of contact is the thin edge of the metal that touches a hook made from the round stock. The actual point of contact is about the size of the period at the end of this sentence. Are you now beginning to get the picture? If you happen to be using an S-shaped hook that swings from both ends rather than being rigidly fixed or welded to the hanging fixture, then you have two weak links. The next weak link is the point of contact between the hanging fixture and your conveyor. The final weak link is the premise in your mind that convinces you that the conveyor is well grounded because all of the supporting steel welded to it is welded to base members of your building that go right into the concrete, thus into the ground.

My friend, you are living in a dream world. Many powder and wet electrostatic systems have been reduced to uselessness because of poor efficiency of systems with poor grounds. Look at it another way. Suppose you know from tests run during the first month of operation of your new powder system that your primary transfer efficiency of virgin powder to part is 70 percent. When your efficiency is subsequently reduced because of poor ground to about half of what it was when you originally tested, what's left? I know you'll recover a part of the powder, but you'll still be increasing your reject rate.

There are many ways to ensure a good ground while your parts are within the powder coating enclosure. The methods are good insurance not only from the standpoint of a good conveyor ground, but also in the name of safety. (It is also necessary for your coating enclosure and accessories to be well grounded.) The ground can be located at the coating enclosure. The two methods I have seen used that seem to work well are shown in Fig. 11.8. One uses a rub bar; the other, copper brushes. The grounded rub bar, usually made of copper, goes the entire length of the coating enclosure. The bar is adjusted so the hanging fixture rubs against it firmly enough to cause a continuous solid contact between the bar and the fixture during its trip through the enclosure.

The wire-brush method uses two copper wire brushes, one mounted on each side of the enclosure and running the length of the enclosure. The portion of the hanging fixture that touches the grounded brushes becomes grounded, if it is clean. If powder is permitted to collect on the brushes, it must, of course, be removed when colors are changed or it could contaminate the next color.

Loading Hangers and Parts on the Conveyor

Depending upon what type of hangers and parts you have, you can load hanging fixtures off-line, then hang the fixture on the moving conveyor as it passes the loading station. Other types of systems permit you to load onto permanently attached hanging fixtures day in and day out. Another system lets the specific hanging fixture be loaded on the line empty; further down the line, parts are hung on the fixture, removed when finished, and replaced with new parts.

Each case is different and depends on what you're hanging. But remember that you need a good, solid, clean contact. And simplicity. I have seen a system which works well at most finishing-line speeds. It's a homemade type of device that uses either a thin sheet of steel or heavy wire. The upper edge is of sawtoothed shape, which enables a hanging fixture to be rapidly dropped into place. These devices are suspended between each conveyor trolley and are attached to the trol-

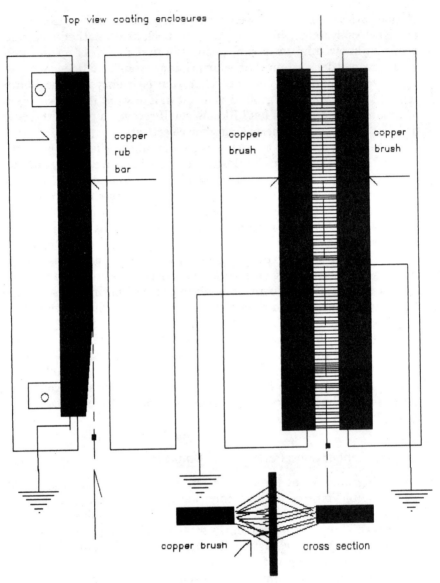

Figure 11.8 Two methods of grounding, one using a copper rub bar; the other, copper brushes.

ley. Most enclosed-track conveyor systems have trolley attachment points on 6- or 12-inch centers. A similar system works well on larger I-beam systems that use swivel pins at each trolley connection.

An I-beam system with a space of 18 inches between trolleys might be used with swivels and some "slop" between the connections to allow for the distance changes in turns. Figure 11.9 gives some ideas as to how these devices work.

The point I would like to bring out is that many times the person loading the part is trying to locate and attach a fixture into a very small hole while the conveyor is moving along at a fast speed. It's like trying to thread a needle with a very shaky hand.

Figure 11.10 illustrates what can happen when hanging fixtures lose hooks or are not completely loaded, and also what can happen when parts fall from fixtures during the finishing process. In all of these situations, the powder-to-part transfer efficiency is affected.

Balancing the Loading

Unless you are fortunate enough to be able to coat the same part day in and day out, you will need to "balance your load" so it is not nec-

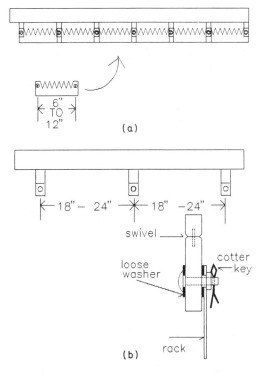

Figure 11.9 (a) The sawtooth pattern makes it easy to hang fixtures while the conveyor is moving. (b) The swivel, loose washer, rack, and cotter key make it easy to negotiate corners at an 18- to 24-inch distance.

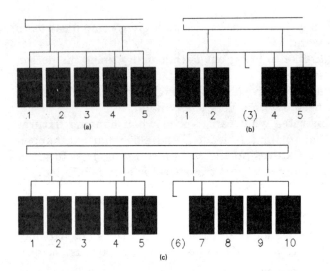

Figure 11.10 Examples of efficiency/inefficiency in hanging. (*a*) There is 100 percent hanging efficiency for the fixture carrying five parts. (*b*) There is 80 percent efficiency for the fixture carrying four parts; a reject is also a possibility for parts 2 and 4, which will get the powder intended for the missing part 3. (*c*) There is 80 percent efficiency for the fixture on the right; again, parts 5 and 7 will get too much powder and could be rejected because of inconsistent mil thickness.

essary to readjust the powder guns every few minutes. When you design your fixtures, try to visualize how they might look as they pass, loaded with parts, in front of your automatic powder spray guns. Try to imagine each hanging fixture occupying the same general space on the conveyor line, regardless of the type of parts. Figure 11.11 shows the "balancing effect" simulated loading of a conveyor with different parts. When assembled, these parts make up an entire toolbox: (1) box, (2) cover, (3) removable shelf, (4) carrying handles, and (5) hinges and hasp.

One of my favorite hook arrangements is seen in Fig. 11.12. A square stock is used as a hook between the part and the fixture. Using the edge of the square stock rather than the flat side tends to let parts slide onto the hook, stripping the edge as they slide, much like two knife blades rubbing against each other. Heavy parts will usually strip coatings as they slide. Light parts will too, if the person doing the hanging has the time to help the part slide down the knife edge of the hook.

In many applications a hook with a V rather than an S as a seat for a part will work. It depends upon the part shape. Some part shapes

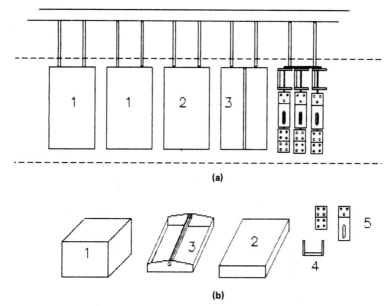

(a)

(b)

Figure 11.11 Balanced loading. (*a*) The area bordered by dotted lines is the balanced area. (*b*) The breakout of the assorted parts. Automatic spray guns spray powder more evenly if the load is balanced "visually" for them. If there were an empty space in the line, parts on either side of it would collect the excess powder, giving poor transfer efficiency.

cause a shadow; other shapes enable the part to be suspended and grounded at two points rather than one point.

Hanging Heavy Loads

Many lines have heavy loads. There are many ways to hang heavy loads. Power/free conveyors offer off-line loading. The load-bar assembly leaves the moving conveyor line to be loaded elsewhere. Several years ago I saw a system which was able to load and unload heavy loads on a straight power conveyor system. The method worked well for the particular job. Figure 11.13 shows how it worked. The product is taken to the staging area on a pallet. The pallet is placed on a floor conveyor. The overhead conveyor takes a drop, the hook on the conveyor engages the loop, and the lower conveyor starts to move. Momentarily, the second hook gets into position and engages the second loop. The hooks and conveyor rail rise up to the previous level, and the loading is complete. The same process works for unloading.

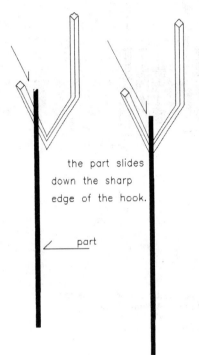

the part slides
down the sharp
edge of the hook.

part

Figure 11.12   A square stock hook.

Cleaning the Fixtures

There are various ways of handling the powder that collects and eventually cures on hanging fixtures. Different situations require the use of different solutions. Some say that hanging fixtures should never be cleaned so long as the point of contact for the part is well grounded. They feel that, as the powder collects on the fixture and is cured, it insulates. Thus it does not collect any more powder, provided the point of contact on the fixture remains clean. But eventually the insulated hanger will develop cracks from multiple trips through the cure oven, and sometimes large pieces of cured powder will break off and fall into collection systems. This could cause a problem. A piece of hard, cured powder could eventually block a passage in the reclaim pumping system, and you'll have to do a lot of troubleshooting to find the blockage. Broken chunks of powder can also cause contamination in the system, or cause speckled parts as color changes are made.

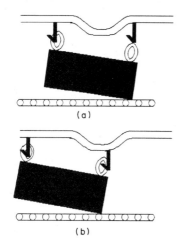

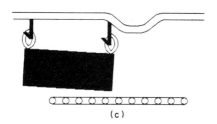

Figure 11.13 Hanging heavy loads. (*a*) The forward hook is attached and the part has started to move along the conveyor. (*b*) Both conveyors operate at the same line speed; the second loop is being picked up by the hook. (*c*) The pickup is complete; the overhead conveyor has the part. The unloading will be done the same way.

So there are those who insist that fixtures need to be stripped frequently to make certain the ground is always good and strong and to prevent contamination. If I were coating a product which was very "touchy" from a cosmetic standpoint, I would rather be conventional and strip the fixtures frequently to avoid the possibility of contamination.

12

Staging

Overview

When you are in the planning stages of your new finishing system, you must include specifications for the loading area of the product or part and consider the logistics of its subsequent unloading. If you are into high part production, the people designated to load the product must feed the new line constantly. Suppose even that with your new system you want to start out by doubling the capacity. On your old system you were coating 8000 pieces an hour for 16 hours a day. Now you want to eliminate the night shift and finish 16,000 pieces an hour. You're going to need to have space at the finishing area to store some of this product to be coated, and you must also be able to remove the finished product from the area. The logistics must be planned now, and not the day the system starts up.

Several years ago we were involved with a startup. We had nothing to do with the planning, just the startup. No advance planning had been made for storing the product to be finished or for removing the finished product from the area. To keep up with production, it was now necessary to load one pallet of product onto the line every 4 minutes. And it was necessary to remove one unloaded finished pallet from the area every 4 minutes. A local forklift rental company reached new profit levels in a few days, as the finishing department struggled to keep up with production. Forklift drivers became more skilled as they quickly loaded and unloaded pallets, racing with each other and concentrating on avoiding collisions. Within a week, the first collision occurred. A lot of money changed hands between employees who had been betting on when the first accident would occur—and between management and the injured employees.

To avoid such problems, you may want to do some serious planning for your new system.

Planning the Loading Area

Consider first the loading area:

- Is there enough storage area for parts?
- How will the parts arrive? By forklift? In boxes? Preloaded on hangers?
- Will the parts be loaded in the delivery area or elsewhere in the plant? Will the parts be loaded by the finishing department? Or will they be loaded at the point of manufacture?
- Who determines how often and by what method colors will be changed? How will this affect loading and unloading?
- How are the logistics to be handled in the case of small parts that must come together for a final large assembly? How will you ensure that parts are loaded and finished in proper sequence and quantities?
- Will you have a power/free conveyor for loading and unloading without pressure?

Figures 12.1 and 12.2 show some methods of storing the product to

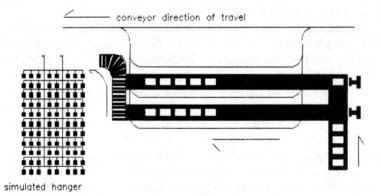

conveyor direction of travel

simulated hanger

Figure 12.1 Storing parts in the staging area. Pallet boxes of parts are loaded onto a "jitterbug" conveyor. As they are needed, the pallets advance to the placed by a ram onto a second line. Empty pallets are removed by a separate line.

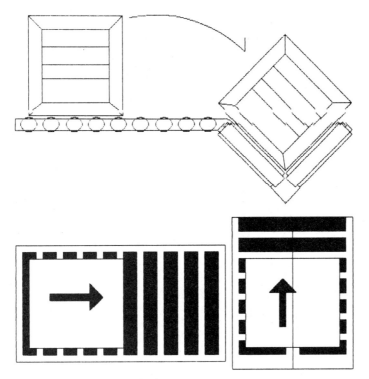

Figure 12.2 Storing parts in the staging area. Small parts are placed in a pallet box, which is loaded onto a roller conveyor, which carries it to a groove where the pallet is tipped, unloaded, and rolled off on a second roller conveyor.

be loaded in the loading area; the method illustrated in Fig. 12.2 is crude at best, but I have seen it work.

Planning the Unloading Area

It costs money to unload parts, place them carefully on pallets with corrugated slip sheets between them, and transport them. There is the matter of labor, material costs, and the possibility of damage. Possibly you are mass-finishing thousands of small parts an hour and placing them in palletized containers for removal to another part of the plant, or into storage, to be used at a later date. It is easier on parts, and more cost-effective, to unload them where and when they are going to be used. It is definitely best to locate the unloading area at the point where the product will be immediately assembled, packaged, and removed.

13

Installation

Overview

The installation of equipment is a profession all its own. Don't get carried away with the idea that you are going to save money by having your own plant people do the installation of your new finishing system, no matter how proud you are of the skills of your mechanics and maintenance staff. There are only two ways to install equipment: Have the people who sell you the equipment install it, or get a professional installation company that has a substantial reputation in the finishing industry to do the job.

Above all, remember this: People who install equipment for a living have experience in doing it. They have done it many times. Your plant people do not install finishing systems every day of the week. I cannot tell you how many times I have seen plant people valiantly trying to install power washers, conveyors, and ovens. I'll admit that they usually end up accomplishing the job, but it takes them at least twice as long as it would professionals—and many phone calls to the equipment manufacturer to ask some simple questions. I'll guarantee that the erection of large pieces of equipment within your plant will require some ingenuity. Experienced installation people can easily supply such ingenuity—and can guarantee that there will be no equipment problems at start-up time. Converted to dollars and cents, it's much cheaper to hire professionals.

Planning the Installation

The installation of a new system needs the same attention and planning as does the buying of the equipment. Your first thought should be for the utilities involved. Get a complete list of all utilities needed for each piece of equipment ordered. You will need to know the total electrical horsepower involved, and in some parts of the country you'll need to establish the exact voltage in advance. How much water will you need? Garden-hose outlets are great for gardens, but one garden hose will not fill those 400-gallon tanks on the new power washer. For the natural gas or propane fuel, you'll need to know total equipment BTUs.

You'll need a dedicated compressed-air system. Do not under any circumstances use a plant compressor. In most instances, your spray equipment will starve for air. Use the plant air compressor for a backup.

You'll need adequate drains. Somehow or other you must get rid of overflow rinse water from your power washer; you'll also have to empty tanks from the various stages. You'll definitely need a floor drain. If there is no drain in the area where you plan to put your system, it is possible and feasible to put a submersible sump below the floor level and pump to your sewer system or holding tank. Although the amount of water discharged is small, it's also a good idea to have a drain for your refrigerated air drier.

I have been involved with projects where, because the left hand didn't know what the right one was doing, we didn't have enough gas and/or electricity in the plant to operate the equipment when we started it up. In these instances, equipment was forced to run on a part-time basis until additional utilities were installed. And even while the equipment was running part-time, labor still had to be paid on a full-time schedule.

Remember to allow for sufficient utility lead time. It's quite likely that electrical wires and gas lines in your neighborhood are not big enough to handle your new equipment. Wires and conduit will need to be set in place, as well as gas lines, air lines, water lines, and any drain lines required. Somewhere in your project you should have a complete layout of your new system. These blueprints should indicate where each utility is needed for each individual piece of equipment. Show these plans to the utilities installers so they can see where they should run the conduit, wires, and pipe. Then, when the equipment is installed, it's a simple matter of making a connection.

Installing the Equipment

Let's take each major component of your new system and examine its probable installation or erection process. Conveyor manufacturers usually have people who can install your conveyor system, which will come in many pieces. Conveyor installation crews hang straight powered conveyor systems very quickly. Because of the second track, power/free systems take a little longer. A conveyor system, of course, will need electricity for its operation; the lubricating and take-up components may need a supply of air pressure. Conveyor switches will be either electric or air-powered; a sophisticated system will need an electrical supply to its programmable controllers. The conveyor line may or may not need safety caging, depending on its elevation.

The power washer will need fuel—natural gas, propane, electricity, or steam. It will definitely need electricity for gas burners, gas trains, and pumps. It will need a water supply and a drain or drains. Depending upon what you bought, and how you bought it, you may need only one connection to the unit for each utility. The freshwater intake and the drain can both be set up with manifolds so that only one connection is needed for each.

Many times the washer manufacturer will build the washer in their plant and ship it in sections by freight truck. In this case, they try and "break" the washer at drain areas.

The oven or ovens, if small enough, may also have been built in advance at a plant and "broken" so they can be loaded onto a freight truck. Ovens need gas and electricity for the burner gas train. Ovens and washers will both require roof openings for exhausts. The roof of your building may be one of those with a guarantee that is valid only so long as no holes are cut in it. In this case, hire a professional who will cut the vents, reseal, and reguarantee the roof. Aside from guarantee considerations, remember that leaking roofs can lead to rejected parts.

Control consoles for the washer, oven, and conveyor come in a variety of packages. Different circumstances require different solutions. If remote-control consoles are needed, rather than consoles mounted on the individual piece of equipment, you'll need to provide wiring at the control cabinets and the equipment.

Powder equipment is not too difficult to erect. Even so, the equipment manufacturer will usually help with the installation simply because they want to make sure their equipment performs properly. Sometimes installers on site for other equipment will set up the powder equipment, provided it's on the plant site when they have completed their other work.

Some Simple Logistics

There are a few important things to look at before truckloads of equipment start to arrive.

1. Who is going to unload the pieces, and with whose equipment?

2. Are the door openings large enough to handle the equipment? Or should you make arrangements to cut a hole in a wall of your building?

3. Is there a clear path from where the equipment will be unloaded to the actual site where it will be erected?

4. Are your unloading areas at ground level, or do you have elevated loading docks? It is easier to remove a 35-foot power washer from a flatbed trailer if the trailer and your plant floor are at the same elevation. Time and money can both be saved at the loading dock.

5. Do you have an unloading crew at the ready? Trucking companies who deliver your equipment like to have people there to unload it when they arrive. If it is necessary for them to wait for an unloading crew to be dispatched from another location, you will probably pay for waiting time. The equipment manufacturer will usually telephone you when the equipment leaves the point of manufacture. This will give you sufficient time to have the unloading people standing by to unload.

6. Do you have the logistics well planned? If you are buying from several suppliers, coordinate the delivery schedules. A major headache can develop when a power washer, an oven, and your new powder coating enclosure all arrive on site at the same exact time.

System Maintenance

Oven Maintenance

Ordinary convective ovens, like other pieces of equipment in your plant, will get dusty and dirty. Occasionally ovens need to be cleaned, especially at the entrance and exit where the air seals seem to collect all the plant dust. Ovens also seem to collect parts that fall from hanging fixtures. If you coat small parts, remove the strays from the oven frequently. If the parts are light enough and small enough, they can get into the duct system, and parts caught in a recirculating air fan can put you out of business for a few days.

The recirculating fan itself usually needs to be greased on occasion. The fan belts as well as those on air seals the oven exhaust should be checked for wear occasionally.

Conveyor Maintenance

Conveyor maintenance can be kept to a minimum by having automatic lubricating equipment installed on your system. But this equipment should be checked and adjusted carefully. Too much lubricant can be as detrimental as too little. Too much will drip on your parts or, worse yet, drip on the floor of your plant. Automatic lubrication systems work very well and need little maintenance if adjusted properly when the system is installed.

Control Cabinet Maintenance

Control cabinets, once set up and with the controls adjusted properly, usually go for a long time without maintenance, provided a little thought is given as to where you locate them. A location away from the heat of the oven and the splash of chemicals is important. Frequently fans are located in the control cabinets to keep cool air circulating through them. This is a great idea, provided the incoming air is filtered to prevent plant dust and stray powder particles from entering the cabinet. Plant dust can cause arcing, and powder can cure or at least melt in the low-100°F range.

Maintenance of Powder Application Equipment

Powder is abrasive. You've heard that before. It travels through the application equipment with air pressure and causes wear on the equipment. Watch for signs of wear in certain areas. This is just good, normal preventive maintenance. To maintain good operating and transfer efficiency on your equipment, check these wear points on a very regular basis. If there is no other way for your application system to eliminate powder fines, then they must be manually removed from the system on a regular basis.

Monitoring the Incoming Compressed-Air Supply

Incoming compressed air should be filtered in order to prevent foreign matter from entering the powder application equipment. The filters should be checked and replaced on occasion.

If your finishing system is one of those which is not shut down at all during daily operation, it's a good idea to protect yourself by equipping it with a dual-intake compressed-air system. Figure 14.1 shows such a system; it's a rather simple operation. The gauge on the upper left shows 80 psi. The one on the upper right shows 80 psi. If the one on the right starts to drop and the one on the left remains the same, it means a reduction of pressure at the filter. Simply adjusting the valves can bring the lower system into operation, allowing you to shut off the upper system while the filter canister is replaced. In this way, there is no need to shut the whole system down.

Power Washer Maintenance

When your washer is installed, you'll be given several demonstrations of how to use it. You should also receive an owner's manual along with the washer. It will explain many of the maintenance procedures you should follow to keep it running.

Expect some nozzle maintenance. Nozzles tend to accumulate particles sent through the system by the pump. Occasionally, particles of rust from the inside of headers and risers will also become lodged in a

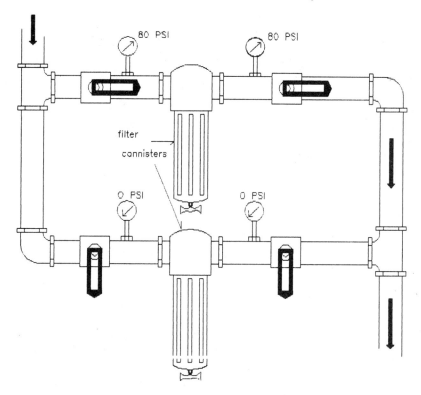

Figure 14.1 A dual-intake compressed-air system. The upper section is the normal operating section and remains so as long as both gauges read 80 psi. But when the gauge on the upper right begins to drop, the upper valves are closed and the lower ones opened. The system then operates through the lower intake while the filter in the upper intake is replaced. The process is reversed when the clean filter is in place, and the upper section becomes operational again.

nozzle opening. Whatever the source of the particle, it must be removed. A good, frequent (weekly) inspection of the nozzles will help keep your washer operating at its peak.

Be prepared for a hard, crusty buildup of chemicals on the outer surface of burner tubing. I have seen some cases of this completely ruin washer operation, sometimes without anyone realizing it was happening. I saw a crusty buildup of about 1 1/2-inch diameter on one washer. No one had ever checked the burner runs. Occasionally pieces of crusted chemicals will break up and go through a washer pump. They may get through the pump, but they will certainly plug up the nozzles.

Remove the scale from the washer at least once a year. Sometimes the cleaning will need to be done more frequently. Buildup on the surface of the heat source is normal. But untended, it will insulate the heat source from the liquid it's supposed to heat, and eventually the scale will fall from the heat source and get into the pipes or risers that supply the chemical to the nozzles. The risers and the nozzles will eventually get plugged up and cut off your chemical source or reduce it to a point where it does you no good.

Burner units on the washer and the fan belts utilized at both the entrance and exit to the washer will need checking on occasion.

No one likes to clean power washers; it's a job that's usually put off every time it's suggested until finally it's forgotten. At no small cost. Let me set a scene for you here. It's based on a true story, disguised only enough to protect the people involved.

The Days of Our Lives

One day—during your busiest production period, of course—your entire morning's production is rejected. Your first reaction is to blame it on Murphy's law. But you need to pass the buck here to someone other than Murphy. You call a fellow from your maintenance department to the scene and ask him to explain what happened. In a frantic effort to keep his job—you're really leaning on him now—he starts the investigation by examining the rejected parts. He immediately notices that the cured powder film is flaking off the parts.

"The powder is at fault," he exclaims, and then relaxes; the pressure is off. You thank him for his quick and intelligent diagnosis. You quickly call the powder company and apply an extraordinary amount of heat to the telephone wires. In an effort to allay your obvious stress, the powder representative suggests several things to be checked immediately and promises to dispatch forthwith her very best troubleshooter; you are, after all, her most valued customer.

When the troubleshooter arrives, he's packing what appears to be a gun. Then you notice that it's only one of those banana-shaped mil-thickness testers. From his belt holster he now draws a crosshatch tester. The whole thing begins to remind you of an old Ronald Reagan late-night TV movie. But all of these quickdraw tactics have only enabled him to find out that the powder does, indeed, fall off in great sheets and right there in front of his very eyes. The troubleshooter deftly picks up a piece of the cured film from the dusty floor, drops his mil-thickness tester, picks up his battery-operated magnifying scope, and professionally examines the dusty flake of cured powder. Straightening up, he smiles and announces that the true problem has been found. He shows you the flake under the 30-power scope. It's obvious that there's contamination on the underside of the film.

"It could be one of several things," the troubleshooter begins. "The chemicals in the washer might need titrating, or perhaps the washer wasn't even turned on, or, most probably, there's something wrong with the electrostatic field of your powder-coating application system."

Your icy stare at the finishing-line foreman, who has of course observed all of this, causes him to immediately produce a record showing that the proper titration was made within the past 2 hours. When the problem with the system surfaced, the titration had been immediately checked again. This second check gave the same results as the first.

Your stress increases by the minute. You suggest the foreman immediately call the chemical salesperson and the idiot who sold you the powder application equipment.

In the meantime, your maintenance man has noticed a very peculiar thing about the washer. Although the gas burner is roaring away in stage 1, the temperature gauge shows the tank to be at ambient temperature. Without batting an eye, or checking to see what the temperature is set at for stage 1, you immediately wheel and fire the foreman, questioning his ancestry as you do so. You almost simultaneously order the maintenance man to cover all bases, to call the gas company, the electrician, and the washer manufacturer and have them get out to the plant on the double.

Having made his calls, the maintenance man returns to the search for the problem. He observes that although the washer gauges are showing sufficient pressure, the nozzles in stage 1 are not spraying liquid as rapidly as they had when the washer was new. Wanting to retain his newly found friendship with you, he calls the water department, and tells them that something has gotten into the water supply and that they had better send an emergency crew out to the plant immediately. Just in case they need help, he also places a call to the plumber.

In the meantime, of course, your ex-foreman has filed a grievance with the shop steward, who arrives at the scene to inform you that unless you rehire the man, the union intends to stage a walkout before the day is over.

Now everyone else begins to arrive. A skirmish breaks out between the powder equipment salesman and the powder troubleshooter. The plumber and gas man discuss union jurisdictional problems, and Murphy's law seems to hold the world in sway.

Our hero (though he doesn't know yet that that's to be his role) now arrives in his Rolls Royce. He immediately identifies himself by giving everyone assembled a ballpoint pen and a stainless steel pocket rule made in Korea. The equipment man eases the tension even more by suggesting that everything be discussed over lunch. You agree that this is a great idea. The chemical man takes this opportunity to casually suggest that the tanks be dumped over the lunch hour. New chemicals, he says, may solve the problem. You agree to that too. So the maintenance man opens the drain valves as the parade of salespersons and tradespeople heads for the closest cocktail lounge.

With the second martini comes an emergency telephone call from the plant. Thinking about the proposed walkout at the plant, you stall on the call and whiff down a third martini before going to the phone. It's your friend, the maintenance man, who informs you that as the tanks were being emptied he noticed a 3-inch coating of a hardened white material caked around all of the burner tubes; there was about 4 inches of the same stuff coating the floor of the tank. Thinking as rapidly as three martinis will let you, you make some fast decisions. You promise the maintenance man a raise on his next check. Then you get him to put the union steward on the phone; you tell the man to tell your ex-foreman that you're hereby rehiring him at a raise in pay. The grateful foreman then gets on the phone. You give him his orders: he's to start chipping this stuff off the burner tubes, clean out the tanks, and check the risers and nozzles. He's then to put the descaling chemicals into the washer, the chemicals which are now on their way to the plant.

Returning to the table, you corner the chemical salesman and tell him to call his company and have the descaling products sent right over to your plant. Now you sit back and order your fourth martini, but not before discharging all of the tradespeople, telling them that the problem has been solved and just to bill your company for the service calls. Well-fortified with martinis, you, the chemical salesman, the powder troubleshooter, and the equipment people order lunch. It's agreed that the chemical salesman will pick up the check.

That night, you tell your wife all about the rough day you put in at the plant. Then you open a beer, turn on the tube, and fall asleep during Monday night football.

15

The Manufacturing of Powder

Overview

Though the powder itself may seem all but insignificant to us, the process by which it is made is truly a miracle. If you ever have the opportunity to watch the process, do so, and you will appreciate what powder manufacturers go through to bring you exactly what you need.

When powder was first manufactured, it was a slow, inefficient, and cumbersome process. Probably in those days powder manufacturers couldn't even envision what would be entailed in an order for 10,000 pounds a week. But as powder sales increased and manufacturing demands grew, powder manufacturers realized that they needed to find ways of making powder faster and more efficiently—and in larger quantities, with extremely accurate color tolerances and repeatable physical properties. These problems were solved when, in the mid-60s, a major resin supplier and Buss, a maker of continuous compounding equipment for the food and plastics industry, jointly developed and patented the technology for continuous processing of powder coatings.

A Three-Stage Process

In this continuous process, the powder goes through three separate stages, during which it takes on three completely different appearances. Figure 15.1 illustrates the process.

The first stage

In the first stage (Fig. 15.1a), the powder formula, which consists of solid resins, fillers, pigments, curing agents, and other ingredients, is

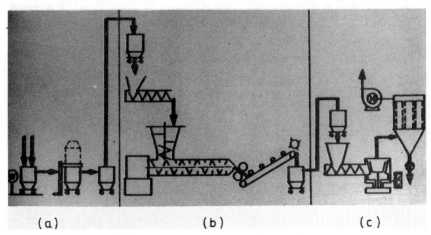

Figure 15.1 The three-stage powder manufacturing process. (*Courtesy of Buss America*)

dry-mixed, or "premixed," in a blender. There are various types of dry mixers, including high-intensity, ribbon, and tumbler blenders. Various ingredients come in different forms. Some of the raw materials are of a crystallike structure, somewhat like rock salt. Others are in a micron-sized powderlike form. The important thing is, these materials must be dry-blended into a uniform state to expedite the second stage.

The second stage

The second stage (Fig. 15.1*b*) is the miracle stage. The dry, blended material is transferred to a volumetric feeder, which meters the premix to an extruder. A metal detector is placed in-line to prevent metal from entering the extruder.

The miracle occurs here in the extruder within a space of approximately 12 inches, when the dry mix is forced through the "cramming feed hopper" and into the kneader. Because of the tremendous mechanical energy released inside the machine the solid resins very rapidly start to melt. A shearing motion is caused as the material is pressed between the screw and the internal pins located on the walls of the extruder. As it is kneaded, the material rapidly rises in temperature. Both the screw within the machine and the internal walls of the machine are temperature-controlled by zoned areas. These separate zones vary from 65 to 220°F. Thus the temperature of the material being kneaded remains at the temperature best suited for that specific material.

The now-molten material is extruded from the kneading machine as a steady output of a semiliquid mass about 1 1/2 inches in diameter.

The mass feeds onto a pair of chilled steel rollers that squeeze it into a solid strip approximately 24 inches wide and about 1/16 inch thick. The strip falls onto a solid, chilled, belted, stainless steel conveyor, which runs at a slight incline. The melted material is cooled after traveling about 16 feet and enters a chamber which crushes it into chips, which are then collected in a suitable container.

The third stage

In the third stage (Fig. 15.1c), the flakes are transferred to a hammermill-type pulverizing machine which grinds them into micron-sized powder particles. The dust created in the process is eliminated, and the usable particles are packed into the suitable shipping containers, similar to the ones you receive and use at your plant.

The Buss Extruder at Work

One of the most popular types of compounding extruder is the PLK 46 Buss America kneader. This kneader is shown with a feed hopper attached in Fig. 15.2. This model has a 12-inch (approximately) working or kneading area throughout. It has a throughput of about 275 pounds of powder an hour. Its larger brothers have capacities of up to 8000 pounds per hour. Figure 15.3 shows a drawing of the screw and the flights which help to shear the molten material. Figure 15.4 shows the

Figure 15.2   Buss kneader with hopper, Model PLK 45. (*Courtesy of Buss America*)

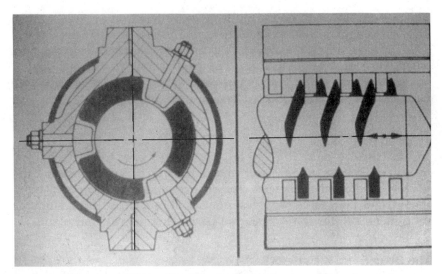

Figure 15.3   Screw and flights of the Buss kneader. (*Courtesy of Buss America*)

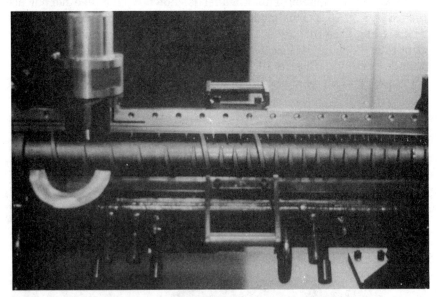

Figure 15.4   Interior of the Buss Model PLK 100. (*Courtesy of Buss America*)

interior of a Buss model PLK 100, which exposes the screw and flights.

Figure 15.5 shows a ribbon of once-molten powder as it leaves the cooling rollers and travels on the chilling conveyor toward the chamber where it will be broken up. Figure 15.6 shows the chilling rollers,

Figure 15.5   Ribbon of chilled powder. (*Courtesy of Buss America*)

Figure 15.6   Cooling rollers, conveyor, and chipper. (*Courtesy of Buss America*)

the chilling stainless steel conveyor, and the device that breaks up the chilled, hardened ribbon of material, which will subsequently be ground into finished powder.

16

Training and Education

Overview

If your company has made a sizable investment in a new finishing system, it would be appalling if the equipment was damaged by employees who were not well trained in its operation. Learning everything there is to know about your new system cannot be done in a day. If that's all of the time you allocate to the training period, you've made a grave mistake.

A plan as complete as the one you made for the design, purchase, and installation of your equipment should be developed for the education and training of your employees.

Powder Training

Your powder company representative will be happy to come in and talk to your people about powder long before the system is installed. Most people employed by powder companies realize that their powder performs its best when used by people who care. If you think it's not important for your finishing-line employees to learn about powder, you're wrong. People who know the facts are better finishing-line employees—and an asset for your company.

If possible, the employees should see powder equipment in operation in another plant. Make certain they see a plant that maintains clean powder areas. It will show them that powder systems are inherently clean, nonodorous areas, where one could lay out a blanket at lunch time and have a picnic. It is important they learn that spraying powder is not like spraying water from a garden hose. They should learn about cure cycles. They should see partially cured parts and learn to distinguish them from fully cured parts. They should learn that if touch-up is needed, it is easy to do. They should learn the value of having a powder finish on your product.

Equipment Training

My basic plan for training employees in equipment operation reads like this:

1. Learn all about powder finishing before the system is running.
2. Learn about the conveyor system.
3. Learn about pretreatment and the power washer.
4. Learn about the dry-off and the cure oven.
5. Learn all about the powder application system.
6. Learn how to clean up properly, with the proper tools.
7. Learn all about general equipment maintenance.

It will take some time to familiarize your staff with each major component of the system. Let the employees scheduled to have the responsibility of each individual piece of equipment learn all they can before the factory people leave your plant.

The conveyor system representative will show your people how easy a good system is to maintain. Lubrication is a must, of course. Automatic lubrication equipment makes maintenance easier. Anyone can hear the alarm a lube system gives when trouble's brewing; but if the equipment is adjusted properly, the alarm will rarely go off anyway. Your people should learn about conveyor-chain take-up systems, stretching chain, conveyor drive, and drive clutches; they should also learn about the complete maintenance of all these components.

Now is a good time for people to learn techniques of fixture hanging on the conveyor and of part hanging on the fixtures. Proper loading should also be reinforced at the washer, oven, and powder application areas. Sometimes unbalanced hanging fixtures can cause problems while traveling through other portions of the system. When an improperly hung fixture or part decides to "grab the wall" (grab the entrance/exit portion of your equipment), something is going to give. The conveyor clutch will go out, or the part or fixture will break, or perhaps the part or fixture will act like a can opener on your equipment.

To get a good finish, you need a good start. To get a good start you need to make certain that at least two people learn how to titrate your aqueous chemical pretreatment system. They must learn not only how to titrate, but they must learn the reasons for titrating. So often, people just learn how to titrate and have no idea of why it is important to titrate. Make certain your employees understand the purpose for each stage of the washer. Good chemical pretreat companies will be very interested in training your people, because a well-trained operator will save them service calls at a later date.

The time to learn about titration is when the power washer or tank system is originally charged with chemicals and water. The trainer can explain how everything starts and stops as the initial charge is being made. Operation of the control console is fairly simple, but explanations must extend to the operation of drains, automatic trickle-fill devices, and pressure gauges. Your employees must also become familiar with overflow adjustments, nozzle adjustments, and nozzle-pressure adjustments.

Around the same time the washer process is being learned, the oven will be installed and balanced. Your employees will have the opportunity to see exactly how an oven is balanced. They should assure themselves at this time that the oven will in fact cure everything they put into it. All in all, they should have a couple of days to learn the washer and oven operation.

Now, when everything else is working properly, the powder application equipment should be started up. You can be in production within a couple of hours. Application start-up people understand their equipment thoroughly. Just make certain the things they take for granted are explained to your employees. Most powder application people will "walk" the employees through the entire process at least once, sometimes twice. Your people should be encouraged to ask questions. Having a recorder to tape all of this won't hurt, since it will give employees the chance to refresh their minds later.

If you will be doing color changing, at least one color change should take place when the factory person is still on site. It would even be preferable to make two changes, the second one being made by the employees alone.

If you did your homework properly, you'll have your cleanup equipment ready to go as well, so that the employees can be trained immediately in cleaning the application area and avoiding the dangers of contamination.

About 30 days after start-up, arrange for a retraining session. Be sure to provide your employees the opportunity to ask questions of the application people now. Many things will have happened during the first 30 days, and your people will now know the right questions to ask. Who knows? They may even ask questions the vendors can't answer. I have seen this happen many times. People working equipment sometimes observe things that vendors never see. Each piece of equipment has its own idiosyncrasies. Problems you have, your neighboring plant may never have seen. Encourage question asking. The more they ask, the more they know, the better equipped your employees will be to do a good job for your company.

Retrain and reeducate your people by sending them to workshops whenever possible. Finishing your part properly is the best sales and marketing tool you have to offer. Woe to the company that puts out a product that loses its finish quickly!

17

Buying Equipment and Materials

Overview

It makes no difference whether you are retrofitting an existing system within your plant or starting out from scratch, remember the "Four Cs" of powder coating as you start to develop your specifications: convey, clean, coat, cure.

You'll need to establish the quantity of your product you'll expect the new system to finish in one shift. You'll need to establish how many lineal feet of conveyor and processing it will take each hour to get the product finished. This will help you establish the approximate length of each individual piece of equipment.

You'll need a basic plan to start with. Some of the things you must consider are given here to be used as a guideline if you happen to be working alone, or with very little help.

1. Basic ideas

   ■ Create a "wish list," the *ideal* plan that includes everything you could possibly want.

   ■ Create a *practical* plan that includes everything you need to do the job properly.

   ■ Create a *utilitarian* plan that includes the equipment management is likely to let you buy.

2. Basic specifications

   ■ Write a basic set of specifications for the pieces you'll need as a minimum.

   ■ Get some estimated costs from vendors.

- Get some estimated costs for installing equipment.
- Estimate the cost of permits.
- Establish your utilities needs. Do you have enough gas, water, electricity, and compressed air in the plant? What is the total horsepower needed for equipment? How many BTUs per hour will be needed? How many gallons of water per hour will be used?
- Estimate the cost of hooking up utilities.
- Verify that your preliminary plans include enough room for the system in the space allocated by management. Inform everyone as to the specific boundaries of the new system.

3. Future projections

- Press for a decision on whether the budget will be adequate for the system you've planned.
- Verify that the specifications you've set forth will indeed do the job properly. If not, establish whether the plan is still feasible. Establish whether, given the budget, additional equipment can be added in a year.
- Accept no unwritten promises (which are cubic feet of empty air). Plan only on what is actually budgeted on paper.
- Withhold payment, or partial payment, on purchases as guarantee of satisfaction.

4. Politics and other variables

- Make sure everyone in the company knows the extent of the plans for the new powder coating system.
- If you're in charge of the project, tie up all the loose ends. If you're only second in command, disengage yourself if necessary from the corporate dreamer who may be in charge. Be sure that the final decision maker holds the same vision you do.
- Take every precaution against failure of the new system since your job may be on the line.

Drawing Up Conveyor Specifications

Chapter 8 will have given you some good ideas on the conveyor system you'll need to do the job within the allocated finishing area.

Conveyor specifications will be based on the type of conveyor selected (enclosed track or I beam) and on the size of the carrier and the package. Figure 17.1 shows the dimensions necessary in calculating

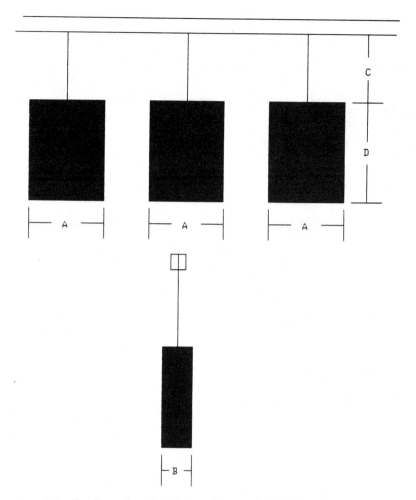

Figure 17.1 The dimensions that have to be established when designing a conveyor system.

the size of your conveyor. The total weight load to be carried on system at any one time can be determined easily, as shown in Fig. 17.2, where A is the weight of your heaviest load, including the weight of the hanging fixture. Beside the weight of the load, you'll also have to establish the horizontal distance encompassed in A. To this add B, the distance between the loads.

Let's draw up the specifications for a conveyor, using Fig. 17.2. Assume your system layout will measure 600 feet. Assume that A will be 2 feet in length and will weigh 60 pounds. Assume that B will be 1/2 foot in length. Then,

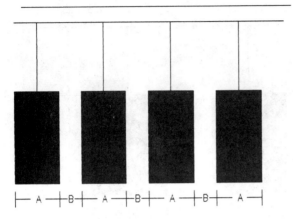

Figure 17.2   Sketch to assist in designing a conveyor system.

$$600 \text{ ft}/2 \text{ } 1/2 \text{ ft} = 240 \text{ packages the size of A}$$

$$240 \text{ packages} \times 60 \text{ lb} = 14{,}400 \text{ lb} = \text{total level load}$$

The 600-foot conveyor length must therefore carry a total load of 14,400 pounds, and

$$14{,}400 \text{ lb}/600 \text{ ft} = 24 \text{ lb-load per foot}$$

Approximate elevations of the conveyor as it passes through the various pieces of equipment should be shown in the sketches which accompany your specification sheet to conveyor suppliers, especially if the supplier will be doing the installation. Decide whether it will be necessary to include safety caging below the conveyor at high elevations.

Purchasing Powder in Large Quantities

The amount of powder purchased will, of course, depend upon your needs. The more powder you purchase, the lower the price generally.

If you are just getting started in powder, and are happy with the results of tests performed by one or more powder suppliers, take a look at the quantity of powder you will need for the next year. Powder manufacturers are able to reduce prices with large-quantity orders, and they will give a price for a one-year commitment.

To help you understand the logistics of all this, let's look at a hypothetical example of a blanket order. Suppose you think your annual requirements will be met by an order of 24,000 pounds, or 2000 pounds a month. You give a purchase order for the 24,000 pounds at a given price per pound. The manufacturer makes a sizable quantity of powder at one time and ships it to a regional warehouse. Once a month, 2000 pounds is shipped to you, and you are billed for it. As the year progresses, you and the powder representative will check your supply occasionally to see if you are, in fact, using the quantities you have ordered.

The powder suppliers want your business and understand that service is part of what you're contracting with them for. They'll make certain you don't run out of powder, unless you suddenly start a lot of overtime or add a second shift and forget to notify them of your change in operations.

Purchasing Powder in Small Quantities

Not all of us purchase powder in large quantities; some companies, particularly powder-coating job shops, use only small quantities. Sometimes you yourself may get a small "private-label" job and need only small quantities of powder. You'll find that some powder manufacturers aren't able to gear down their production quantities; their production machinery just cannot make small quantities. But, I'm happy to say that there are companies who can accommodate you with small quantities when and if you need them. Of course, the price will be higher per pound, but you should be able to adjust your product price to cover the extra cost.

Deciding on the Power Washer

If you have read Chap. 4 on power washers and pretreatment chemicals, you'll already have a good idea of your needs in a washer. You'll also have a good idea of how long your washer should be, having established each stage length. The tests tell you that it will take 1½ minutes to clean the parts of all soils: 1½ minutes times your projected line speed gives you the length of the first stage. The distance between stages is predicated on the overall length of your product in the direction of conveyor travel.

You will also need to know how many pounds of product, hanging fixtures, and conveyor chain will be going through the washer every hour. You will need to establish the silhouette, package, or profile size, so the openings can be designed properly. Decide now whether you'll be placing your washer in a pit so that the dry-off oven can be butted against the washer, or whether they'll both be stand-alone units.

Washer manufacturers will tell you what they usually supply as standard equipment. Make certain you get what you feel is necessary for good production.

Deciding on the Ovens

You know by now whether or not your dry-off and cure ovens will be together as a single oven or separate units. You will also know if the dry-off is to be butted to the power washer. You will need to furnish your vendor with weight per hour through the oven, plus the profile, silhouette, general shape, and size of the part or product.

You will need to know the cure cycle of the powder, the heat-up time of your parts, and the general size or shape you need in an oven. You should know by now how many turns the conveyor will be taking as it passes through the oven.

Deciding on Powder Application Equipment

Chapter 6 will have given you many ideas on powder coating equipment. You will need to decide on the size of the openings at each end of the coating enclosure. You should know something about colors—the various quantities you'll be using, the frequency of color change, the powder system that will work best for your particular production. Your only decision now is how much money and space to invest in the equipment.

Deciding on the Location of the Control Cabinets

Now that you have a basic plan, you should know where your product is to be loaded and unloaded. Will the operators be able to see the control cabinets of the washer and oven? Will the control cabinets have alarms so someone will be alerted if there is an equipment problem? There is nothing more embarrassing or costly than to have a few hours' worth of production come through partially cured, or imperfectly cured, because no one noticed an equipment failure. It's important that someone be responsible for monitoring equipment frequently. It is easy to do this if the control cabinets are close together and within someone's visual range all of the time.

Ordering Pretreatment Chemicals

Your complete pretreat process should be established by now. Someone helped educate you and ran all of the chemical tests with you. You should now show this person how much you value the service given to date by giving him or her the initial order and the assurance that you'll be depending on continued service and training as the system begins operation.

A good pretreatment supplier will make certain that your system starts up properly and that your people are well trained. The supplier knows that future orders will be predicated upon how well the chemicals perform during the day-to-day operations.

Having It All Come Together

Experienced hanging fixture suppliers can ship on good schedules, but from experience I can tell you some sad stories. I have been at start-ups where the hanging fixtures had not yet arrived. In fact, it was necessary for the equipment start-up people to return at a later date to train employees. Poor management and poor coordination can betray the best intentions.

Your hanging fixtures must be at the plant and ready to go so oven-load adjustments can be completed when the oven is first started up. The complete system should be in production within a day or so after everything has been started up and tested.

System Transfer Efficiency

Overview

"Transfer efficiency" is an expression you will hear many times in the finishing industry. Like any other industry, the finishing industry has its own jargon and "buzzwords." They seem to belong to us. Sometimes I have wondered if maybe we're simply trying to impress people with our terminology. Certainly you've been involved in a conversation with people who use nothing but buzzwords. Be honest now. Don't you usually just shake your head "yes," even if you don't know what they're saying, just so you can stay in the conversation and not appear ignorant?

"Transfer efficiency" is one of those much-used, much-maligned phrases. If you make the mistake of asking the speaker, "Just how high is the transfer efficiency?" you will usually get the raised-eyebrow treatment and the curt reply, "*Very* high!" But when you get your system into the plant and operating, it may dawn on you that the transfer of the powder is not as high as you expected it to be. Examine the entire system very carefully. No one item is responsible all by itself for transfer efficiency. Excellent transfer efficiency is a combined result of all functions of your system. All the components—the various pieces of equipment, the powder, the hanging fixtures, the part design and part shape, the exact location of the system within your plant, the employees themselves—are all factors in the system's transfer efficiency. Any part of your system which isn't running at its highest peak of performance is going to help drag your efficiency down.

Remember: Bringing high transfer efficiency to your production line will result in important cost savings to you.

Transfer Efficiency and Powder Application Equipment

The electrostatic output of your guns is variable. You will find that by adjusting the output for different parts you will get better efficiency in the spraying, hence better overall transfer efficiency. The quantity of powder expelled from your guns must be controlled carefully. The voltage, air pressure, gun position, all must be taken into consideration when thinking about transfer efficiency. Remember, you only need enough powder to coat your parts, not the entire interior of the coating enclosure. The "assisting" air need only be enough to get the powder particles to your parts, no further. Powder guns must be blown out occasionally to keep them working at their peak. If you're monitoring them closely, you can usually tell when gun patterns change. Blowing off the tips of the guns or blowing out the nozzles will usually get you back onto the right track.

As I have mentioned before, gun manufacturers make several types of nozzles and tips. Try to match the tip or nozzle with your parts. You'll find some tips work better than others. Eventually, nozzles, electrostatic tips, hoses, and powder pumps start to wear. As they do, your efficiency will change again.

Ancillary equipment such as sieves, feed hoppers, and reclaim systems must also work at their peak efficiency. They must be kept clean and free from impact fusion points as much as possible.

The company that manufactured your powder equipment spent a lot of money on their service manuals. Reading and understanding these manuals will help you get better efficiency from your equipment.

Transfer Efficiency and Proper Hanging Techniques

The spraying of large flat panels will usually give you excellent transfer efficiency. If your equipment is adjusted and working properly, the charged powder propelled from the gun will, by electrostatic attraction, form a fine film on the clean part suspended from a clean hanging fixture. All of which are properly grounded, of course.

But not everyone is fortunate enough to be constantly spraying flat parts. Most of us have wire goods, mixed shapes, welded intersections, holes, Faraday cage areas.... It's these realities you'll have to have the patience, and the skills, for coping with.

Remember, the automatic spray guns are blind. Your hanging fixture must present the parts properly to the guns, must give them plenty of product to coat.

Good clean fixtures + plenty of good clean product =

excellent transfer efficiency

Transfer Efficiency and System Location

Plan your new system layout carefully. The location of the coating enclosure itself can be critical. For example, if the coating enclosure or reclaim system is sitting next to your oven, it may be subjected to high heat. In fact, if your enclosure gets its air directly from hot air coming from your oven, the powder could start to melt. A coating enclosure located in line with the prevailing winds from a loading dock door could also mean trouble.

Under normal circumstances, overspray powder will stay within the confines of your enclosure. But it's possible to have air currents strong enough to overcome the delicate balance imposed by the reclaim system and redirect powder throughout your plant.

Transfer Efficiency and Powder Type

The particle size, spraying temperature, age, and composition of the powder will all affect transfer efficiency. Powder particles come in all sizes and shapes. That in itself will make a difference in the charging rate. Then there is the quality of reclaimed powder, whose particles may have changed in size, and hence charging capacity, from that of the original particle.

Reclaim systems do not generally enhance the transfer efficiency of an oversprayed powder particle. However, newer systems do handle the powder very carefully.

Transfer Efficiency and Finishing
Department Employees

Sometimes management gives no thought to the hiring of the people who will operate their very expensive finishing system. That doesn't make sense. It's been my experience that not everyone likes to work around finishing lines. Some people enjoy it, some people don't. Good, well-operating finishing systems are usually staffed by a team of interested people.

That's not to say that a careless, ambivalent worker cannot be turned into someone who enjoys the job. Several years ago a client changed from a wet-coating system to a powder system. The management assumed that their people would eventually learn the difference in wet, nonelectrostatic spraying and electrostatic powder spraying.

The people were trained and retrained, but they continually went back to using the old wet style of multiple strokes, wasting powder by the box. We eventually suggested to management that they switch the sprayers with the people who loaded and unloaded the line, people who had never sprayed but who had attended all of the training sessions and were willing to give it a try. The switch was made and the system immediately started to increase in efficiency. Management soon increased the wages of the new sprayers while maintaining the wage of the old sprayers; now loaders were earning as much money as sprayers. In time, the old sprayers asked to be moved to touch-up spraying on a part-time basis, and old and new sprayers now work together as a team, trading off jobs, alternately spraying, loading, and unloading.

Calculating Transfer Efficiency

Let's look at how people figure transfer efficiency in a finishing system. Let's set up a hypothetical case; let's say that:

1. You have 60 percent transfer efficiency from your guns to your product; that is, 40 percent of the powder goes into the reclaim system.

2. Being partially plugged, your powder guns are spraying at 80 percent efficiency; a couple of them have developed impact fusion points in the interior of the spray nozzles.

3. You are using much more assist air than you need, giving you a transfer efficiency here of 70 percent.

4. Your reclaim system is reclaiming powder at 80 percent efficiency because you have put off the replacement of your cartridge filters.

5. Your parts are clean, your hangers have just been stripped, and they are running at about 95 percent efficiency.

Now let's figure the efficiency of your system:

Item	1		2		3		4		5		Overall
Efficiency,%	60	+	80	+	70	+	80	+	95	=	385/5 = 77

By our calculations the overall transfer efficiency of your system is 77 percent. Any way you look at it, the finishing area is your profit center, and it must run well all of the time.

Index

ABOUT THE AUTHOR

With more than twenty years of experience designing, selling, installing, and starting up complete powder coating systems, Bill Lehr is one of the foremost experts in the field. As an independent consultant specializing in troubleshooting and designing systems, he conducts in-plant training seminars around the country and all seminars on the subject for the Society of Manufacturing Engineers. Mr. Lehr is widely credited with helping to popularize powder coating technology in the United States.